Robson S. Rocha
Marcos R. V. Lanza
Rodnei Bertazzoli

Electrosynthesis of alcohols using gas diffusion electrodes

Robson S. Rocha
Marcos R. V. Lanza
Rodnei Bertazzoli

Electrosynthesis of alcohols using gas diffusion electrodes

Electrochemical technology for environmental protection

ScienciaScripts

Imprint
Any brand names and product names mentioned in this book are subject to trademark, brand or patent protection and are trademarks or registered trademarks of their respective holders. The use of brand names, product names, common names, trade names, product descriptions etc. even without a particular marking in this work is in no way to be construed to mean that such names may be regarded as unrestricted in respect of trademark and brand protection legislation and could thus be used by anyone.

Cover image: www.ingimage.com

This book is a translation from the original published under ISBN 978-613-9-68064-1.

Publisher:
Sciencia Scripts
is a trademark of
Dodo Books Indian Ocean Ltd. and OmniScriptum S.R.L publishing group

120 High Road, East Finchley, London, N2 9ED, United Kingdom
Str. Armeneasca 28/1, office 1, Chisinau MD-2012, Republic of Moldova, Europe
Printed at: see last page
ISBN: 978-620-8-19289-1

SUMMARY

1 Introduction

Industry is constantly looking for increasingly economical, cleaner and less energy-intensive production processes, replacing reagents and using more efficient catalysts. In the quest to improve production processes, several research projects are based on the development of new catalysts, the choice of lower-cost raw materials that can minimize waste and improve the energy efficiency of the process. Despite these efforts, many processes in the chemical industry still show low efficiency, a high rate of waste and by-product generation, rapid deactivation of catalysts, high pressure values and high temperatures.

This work will present the results obtained in the electrochemical synthesis of methanol and ethylene glycol under mild conditions of pH, pressure and room temperature.

1.1 Conventional synthesis of ethylene glycol

Ethylene glycol is an organic compound with wide applications in various sectors of the chemical industry. Although the best known applications are as an anti-freeze liquid or in the composition of hydraulic fluids, the largest consumption of this compound is in the manufacture of saturated and unsaturated polyester resins, as well as polyurethanes. Among the best-known polyesters are polybutylene terephthalate (PBT), polycarbonate (PC) and polyethylene terephthalate (PET), used in the manufacture of packaging for fizzy drinks. Depending on the application, the latter is also called Terylene, used in the manufacture of clothing and garments.

In Brazil, the biggest consumers of ethylene glycol are polyester resin producers, including Ara Ashland (Ara Quimica S.A.), Cray Valley (Huntchinson do Brasil S.A.), Cromitec Resinas do nordeste S.A., Cytec Especialidades para Superficies Ltda, Denver Polymers Ind. Com. de Produtos Quimicos Ltda., DPV Produtos Quimicos Ltda., Dupont S.A., Elekeiroz S.A., Embrapol Ind. Com. de Polimeros Ltda., ICI Packaging Coatings Ltda., Reichhold do Brasil Ltda., Renner Sayerlack S.A. and WEG Ind. Quimicas S.A. (ABIQUIM, 2006).

Ethylene glycol, or ethane-1,2-diol, is produced by reacting ethylene oxide, or 1,2-epoxyethane, with water in the presence of a catalyst, such as dilute sulfuric acid, at a temperature of 60° C, according to Equation 1.

$$\underset{CH_2-CH_2}{\overset{O}{\wedge}} + H_2O \longrightarrow HO\text{-}CH_2\text{-}CH_2\text{-}OH \qquad (1)$$

Although the reaction takes place under mild pH and temperature conditions, the storage and handling of ethylene oxide is always a cause for concern due to the reactivity of the compound, which is also highly flammable and explosive when in contact with oxygen.

The problems associated with the use of ethylene oxide as a raw material in the production of ethylene glycol lie in the safety issues involved in its manufacturing process and, perhaps for this reason, the small number of manufacturers. For example, Brazil has only one manufacturer of ethylene oxide, Oxiteno do Brasil S.A., with a production capacity of 312,000 tons/year at its Mauà-SP and Camaçari-BA plants. This company also produces ethylene glycol with a production capacity of 310,000 tons/year (ABIQUIM, 2006).

Currently, all the ethylene oxide consumed in the world is produced through the reaction between ethylene and oxygen, at 200-300° C and a pressure of 15 atm, catalyzed by silver, according to Equation 2.

$$CH_2{=}CH_2 + \frac{1}{2} O_2 \rightarrow C_2H_4O \qquad (2)$$

Although C_2H_4O is also the formula for the acetaldehyde isomer, the ethylene oxide formed is a crown ether, as shown in equation 1.

The yield of the exothermic reaction, represented by equation 2, is 70-80% as long as the temperature does not exceed 300° C. The most important side reaction is the combustion of ethylene oxide to carbon dioxide and water, i.e.,

$$CH_2{=}CH_2 + 3\,O_2 \rightarrow 2CO_2 + H_2O \qquad (3)$$

Another reaction also contributes to the reduction in yield, which is the reaction of the ethylene oxide formed with oxygen:

$$C_2H_4O + 5/2\,O_2 \rightarrow 2CO_2 + 2H_2O \qquad (4)$$

The reactions represented by equations 3 and 4 cause the temperature to rise and can put the process out of control.

1.2 Conventional Methanol Synthesis

Methanol is an organic compound belonging to the group of alcohols, with a carbon atom attached to a hydroxyl group (-OH) and an average molar mass of 32.04 g mol^{-1} . Methanol is miscible in water (it forms a racemic mixture) with a density of 0.791 g mL^{-1} , has a boiling point of 64.6 °C and can be purchased on the market at various levels of purity (CAS Number 67-56-1), the structure of methanol can be seen in Figure 1 (Sigma-Aldrich, 2008).

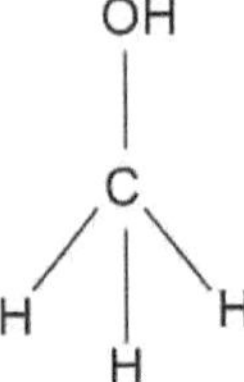

Figure 1 - Structure of methanol

Methanol is widely used in organic syntheses such as the production of formaldehyde, methyl chloride, acetic acid, methylamines, methacrylates, dimethyl phthalate, methyl salicylates, polyester fibers and methyl mercaptans. It is also used for liquid phase extractions, as a solvent in various chemical processes, in gasoline refining and in the refining of certain heating oils. In Brazil, the largest producers of methanol are located in the state of São Paulo, such as Copenor, through Metanor S.A., Fosfertil and Vicunha Têxtil (Abiquim, 2008 and Metanor, 2008).

Methanol is produced industrially from a synthesis gas, where methane (mainly from natural gas) is combined with hydrogen under a pressure of 35 MPa and a temperature of 400 °C. This process of synthesizing methanol requires total control of the system, because in the event of major changes in the temperature and pressure conditions, the yield of methanol synthesis decreases greatly, forming other undesirable products such as CO_2 and, to a lesser extent, formaldehydes (Metanor, 2008 and Anthony *et al*, 2005).

The synthesis of methanol from methane has remained practically unchanged since it was developed by BASF in 1923. Minor updates to the process have been made over the years, mainly in the development and updating of the catalysts used (Khokhar *et al*, 2009).

The synthesis process takes place in two stages, the first of which is the highly endothermic production of synthesis gas (Equation 5), its main limitation being its energy cost (Indarto, 2008).

$$CH_4 + H_2O \rightarrow CO + 3H_2 \qquad \Delta H^\circ = 206 \text{ kJ mol}^{-1} \qquad (5)$$

$$CO + 2H_2 \rightarrow CH_3OH \qquad \Delta H^\circ = -90{,}6 \text{ kJ mol}^{-1} \qquad (6)$$

$$CO + H_2O \rightarrow CO_2 + H_2 \qquad \Delta H^\circ = -41{,}2 \text{ kJ mol}^{-1} \qquad (7)$$

$$CO_2 + 3H_2 \rightarrow CH_3OH + H_2O \qquad \Delta H^\circ = -49{,}5 \text{ kJ mol}^{-1} \qquad (8)$$

The second stage (Equations 6 to 8) is an exothermic process and requires copper as a catalyst. The synthesis reactor is usually operated at high temperature and high pressure to achieve high yields and selectivity for the methanol formation reaction. However, these conditions tend to reduce the lifetime of the catalyst due to the sintering of the metal on the catalyst surface, reducing the selectivity of the process (Indarto, 2008).

Another point to consider in the process of catalytic oxidation of methane to methanol is the purity of the synthesis gas (CO and H_2). Natural gas, an important source of methane, has H_2S as an impurity, which in the presence of catalysts, high pressure and high temperatures is converted to sulphur oxide; Sulphur poisoning of the catalyst occurs due to the very strong bond between this compound and the active sites of the catalyst, thus preventing the adsorption of molecules, directly affecting the selectivity of the process (Khokhar et al, 2009 and Indarto, 2008).

1.3 Electrosynthesis of methanol and ethylene glycol

Looking at synthesis processes using gaseous compounds, such as the oxidation of methane and ethylene, one can see the need for high working pressures even when working in the liquid phase. These specific requirements for the reaction environment in the oxidation of low-chain gaseous compounds show the need to develop processes using new technologies. Within this context, electrochemical techniques have emerged as an alternative, where the oxidation or reduction reactions take place in an aqueous medium and without the need for high pressures and temperatures, making the synthesis processes simpler and cheaper. In view of the severe conditions of conventional synthesis of both methanol and ethylene glycol, this work will consider the use of electrochemical processes for the synthesis of both. Electrochemical processes are cleaner, less aggressive and operate under milder pressure, temperature and pH conditions than traditional processes.

The electrosynthesis of methanol will be carried out by the oxidation of methane, and of ethylene glycol by the oxidation of ethylene gas, both in an aqueous medium. However, electrochemical processes require the reactants to be dissolved in the aqueous phase or in a support electrolyte. This requirement can be a limiting factor in the reaction due to the low solubility of gases in aqueous solutions. Gaseous reagents depend on the transfer of molecules from the gaseous medium to the aqueous medium, directly interfering with the solubility of the compounds. For example, the solubility of methane is 0.03308 cm^3 per mL of water at 20°C and ethylene is 0.226 cm^3 per mL of water at 0°C, both at 1 atm (Gama Gases, 2008).

Faced with this limitation, the use of gas diffusion electrodes (GDEs) may be the alternative to maximize the supply of both methane and ethylene oxide to the reaction medium. Under pressure, these gaseous reagents can percolate through the structure of an EDG to react at the electrode/solution interface in sufficient quantity to meet the desired reaction rate.

Another obstacle to overcome is that the electrosynthesis of both compounds will occur through an oxidation process. In other words, the EDG will be the anode of an electrolysis cell, which requires electrochemically inert, chemically and physically stable materials under conditions of polarization at high anode current densities.

Figure 2 shows the possible route for the electroxidation of methane to methanol. The methane will be kept under pressure on the opposite side of the solution. The anodic discharge of the water will generate the oxygen responsible for the oxidation.

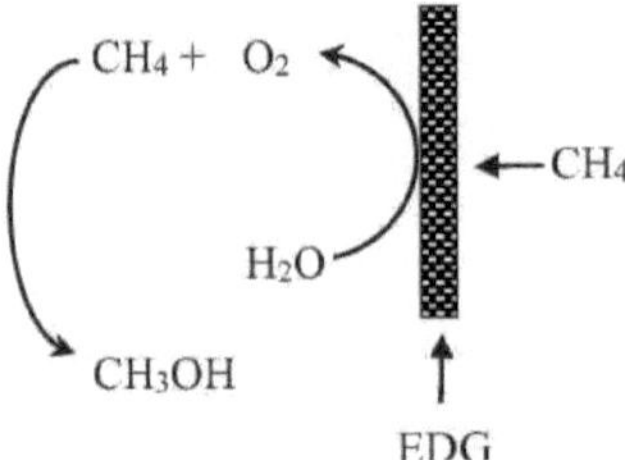

Figura 2 - Scheme of the methanol electrosynthesis process at a gas diffusion electrode.

Similarly, ethylene will be kept under pressure on the opposite side of the solution. At the electrode/solution interface, ethylene will be oxidized to ethylene oxide after the water is discharged. In an aqueous medium, ethylene oxide is transformed into ethylene glycol. The process is shown in Figure 3.

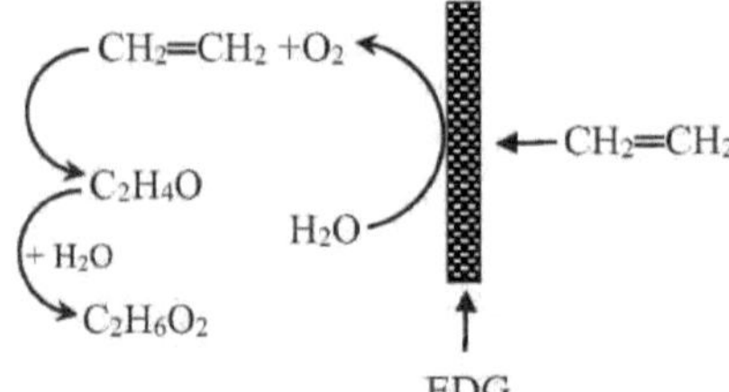

Figura 3 - Schematic of the electrosynthesis process of ethylene oxide/ethylene glycol on an ethylene diffusion electrode.

As such, there are two challenges facing the development of this project, namely the development of porous and inert electrodes at high anodic current densities, which can function as gas diffusion electrodes, and the development of the oxidative electrosynthesis processes themselves.

1.4 Thermal Oxide Electrodes

When it comes to electrode materials that are inert at high anode current densities, the first materials to be considered are dimensionally stable electrodes. These are oxide-coated titanium electrodes, based on RuO_2, IrO_2, TiO_2/ RuO_2, Ta_2O_5/IrO_2, which were popularized in academic research in the 1990s for the oxidation of organic compounds (De Faria et al, 1992; Comninellis, 1994; Foti and Comminellis 1999; Simond et al 1997). The reasons for its success in the chlorine and soda industry, particularly TiO2/RuO2 (70/30 mol%), was reported by Trasatti (Trasatti, 2000).

In the electrosynthesis of methanol and ethylene glycol, the proposed route aims to take advantage of a unique characteristic of this type of electrode, which is the adsorption of atomic and radical species, intermediates in the oxygen evolution reaction. The anodic discharge of water on these oxide electrodes (MO_x) begins with the formation of hydroxyl radicals (OH-) which physically adsorb on the surface, according to the equation:

$$MO_x + H_2O \rightarrow MO_x(OH^{\bullet}) + H^+ + e^- \qquad (9)$$

Mixtures of oxides such as TiO2/RuO2 and Ta2O5/IrO2 have a high density of oxygen vacancies and the radical species adsorbs chemically, forming superior oxides according to:

$$MO_x(OH^{\bullet}) \rightarrow MO_{x+1} + H^+ + e^- \qquad (10)$$

This atomic oxygen would then be released to react with the reactant gases, methane and ethylene, forming the desired products:

$$MO_{x+1} + CH_4 \rightarrow MO_x + CH_3OH \qquad (11)$$

$$MO_{x+1} + C_2H_4 \rightarrow MO_x + C_2H_4O \qquad (12)$$

In the latter, the reaction continues in an aqueous medium, forming ethylene glycol.

The strategy of this work will be to produce the thermal oxides in powder form for pressing and production of the gas diffusion electrodes.

1.5 Objectives

The project's main objectives are:

1- Establish an electrochemical route for the production of methanol and ethylene glycol. For methanol, the direct oxidation of methane will be the main route. For ethylene glycol, ethylene gas is the reagent that will be oxidized to ethylene oxide which, in an aqueous medium, will be converted to ethylene glycol.

2- Develop and characterize electrochemically inert gas diffusion electrodes under electrolysis conditions;

3- Incorporate catalysts into the EDG mass for comparative efficiency and selectivity tests for the products of interest.

4- Quantify the products formed and identify the by-products using gas chromatography and mass spectrometry. Evaluate the electrical efficiency of the electrochemical reactions.

To develop the electrochemical route for electrosynthesis processes, gas diffusion type electrodes (GDEs) will be used, consisting of metal oxide powders obtained by thermal decomposition. The oxides will be pressed with a hydrophobic component and hot sintered in the form of discs to be used as electrodes. The EDG will be pressurized on one side with the working gas and on the other side the electrode will be in contact with the electrolyte.

EDG will be used in the electrochemical synthesis of methanol and ethylene glycol. The process will be evaluated by detecting and quantifying the by-products formed depending on the experimental conditions, such as cell potential, reactant gas flow and the nature of the catalyst incorporated into the EDG.

2 LITERATURE REVIEW

This chapter presents a bibliographical review of the various processes for oxidizing gaseous organic compounds.

2.1 Metal Oxides

In 1994, COMNINELLIS studied the oxidation mechanism of organic compounds with the simultaneous evolution of oxygen in oxide anodes. The author determined that H2O in acidic solution produces hydroxyl radical adsorbed on the oxide anode (MOx) forming MOx(OH), the adsorbed hydroxyl radical can react with the oxygen present in the oxide, forming a higher oxide, MOx+1. According to the author, the two forms of the oxides take part in the selective oxidation of organic compounds, with MOx(OH) taking part in a total oxidation reaction of the organic, forming CO2 and MOx, while MOx+1 takes part in a partial oxidation reaction, forming MOx and the organic in its oxidized form. The author concludes that SnO2 favors the complete combustion of organic compounds and that IrO2 and Pt favor the partial selective oxidation of organic compounds (Comninellis, 1994).

In 2002, PELEGRINO and colleagues studied a laser calcination method for the preparation of DSA-type oxide electrodes®. The results showed the viability of the laser calcination process. When 15W and 35 mm s-1 were used, the appearance of oxidation and reduction peaks associated with the ruthenium redox pairs, Ru /Ru^{4+3+} and Ru^{6+}/$Ru4^{+}$, was observed in cyclic voltammetry. The authors presented results of the variation of the anodic and cathodic charge ratios, where the charge ratio reached approximately 1 during 100 cycles. The authors also evaluated the oxygen release reaction, where linear voltammetry showed an increase in current of approximately 1.37 V vs. ECS, an increase which the authors associated with the oxygen release reaction (Pelegrino *et al,* 2002).

In 2006, MIWA and colleagues studied the electrochemical treatment of the pesticide carbaryl using three different Ti/Ru, Ti/Ru/Sn and Ti/Ir anodes, different current densities and different support electrolytes. The results showed that the three electrodes had a low level of total organic carbon removal in H2SO4, reaching 30% removal at the highest current density studied, indicating, according to the authors, that the removal of organic load does not necessarily depend on the potential applied. The authors studied the addition of NaCl to the electrolyte and the results showed that changing the electrolyte improved the results, reducing the concentration of carbaryl and removing 99.72% of the organic load at a current density of 20 mA cm^{-2} on the Ti/Ir electrode. The results presented by the authors showed that the oxidation of organic matter not only depends on the current density applied, but also on the electrolyte used and the electrode chosen for electrochemical degradation (Miwa *et al*, 2006).

2.2 Gas Diffusion Electrodes

In 2006, GHARIBI and colleagues studied the process of manufacturing a gas diffusion electrode and its modification with (I) polyaniline (PANI) doped with trifluoromethane sulfonic acid and (II) Nafion solution, for use in fuel cells. The authors concluded that the addition of PANI to the gaseous diffusion electrodes improved the results obtained to the point of making it possible to reduce the use of platinum as a catalyst for this type of EDG application. The addition of PANI as a conductive polymer led to a reduction in the electrode's polarization resistance and also gave the electrode's surface a fibrous characteristic, providing a greater active area and a better bond between the catalytic particles. According to the authors, the use of EDG with PANI presented a lower overpotential, minimizing the ohmic drop and the limitation by mass transport (Gharibi et al, 2006).

In 2007, FORTI and colleagues studied the electrogeneration of H2O2 in gas diffusion electrodes catalyzed with 2-ethylanthraquinone (EAQ) using pressurized O2 at 0.2 Bar as the working gas. The results showed that the rate of H2O2 electrogeneration was improved in the presence of the organic redox catalyst (EAQ) and the overpotential for the oxygen reduction reaction was shifted to more positive potentials. The authors concluded that H2O2 electrogeneration follows pseudo-zero-order kinetics and that the addition of EAQ increased H2O2 concentrations by 30% and shifted the potential by 400 mV towards more positive potentials (Forti *et al*, 2007).

In 2008, SOEHN and colleagues studied the development of gas diffusion electrodes (GDEs) built using two techniques: (I) growth of carbon nanotubes on a carbon fabric and (II) adsorption onto the carbon fabric from a solution of catalysts. In analyzing the results, the authors determined that the surface area of the electrode increased with the growth of the nanotubes. The process of preparing the EDG by adsorbing the catalysts promoted wettability characteristics and better particle distribution, with the deposited Pt reaching a size of between 2-5 nm. The performance of the gas diffusion electrode at high temperatures was tested in a phosphoric acid fuel cell, reaching 220 mW cm^{-2} at 150°C (Soehn *et al*, 2008).

2.3 Oxidation of Gaseous Organic Compounds

In 1995, FOULDS and GRAY studied the main factors influencing the partial oxidation of methane in the gas phase. With regard to the temperature of the reactor, the authors determined that there are two temperature values that should be considered in this type of experiment: 1) the temperature of the reaction mixture (methane and oxygen) and 2) the working temperature of the reactor, but most of the studies cited by the authors do not control the temperature of the reaction mixture, but only the temperature of the reactor. The authors also determined that in most of the works cited, the working temperature is between 300 and 500°C and that conversion is low up to the critical temperature (specific to each system), after which the conversion values increase rapidly (Foulds and Gray, 1995).

The authors also reviewed the influence of reactor pressure on reactions involving methane oxidation and identified conflicting trends. Some authors described an increase in the selectivity of methane oxidation with increasing pressure, while others cited the independence of pressure on the selectivity of methane oxidation. Another point reported by the authors was the tendency for the working temperature to decrease with increasing methane pressure, with the increase from 1.5 to 3.0 MPa there was a decrease from 420°C to 380°C and when the pressure was increased to 5.0 MPa the temperature dropped to 370°C (Foulds and Gray, 1995).

In 1995, WANG and OTSUKA studied the kinetics of the methane oxidation reaction, the characterization of the catalyst and the influence of hydrogen on the reaction mechanism. The catalyst used was FePO4 and was prepared by calcination at 823 K for 5 hours in an air-rich atmosphere. The experiments were carried out in a fixed-bed flow reactor at ambient pressure and at various temperatures (Foulds and Gray, 1995).

In the first experiments, the authors studied the influence of the amount of hydrogen on the oxidation results. The results showed that the absence of hydrogen leads to a low methane conversion rate (less than 0.1%), which leads to selectivity for the formation of formaldehyde (40% selectivity) and the non-formation of methanol; an increase in the amount of hydrogen leads to an increase in the methane conversion rate: 8.4 kPa of H2 converts approximately 0.2% and 50.7 kPa of H2 converts 0.5%. Increasing the amount of H2 also increased the selectivity of methanol formation, reaching selectivities of over 20% at the highest H2 pressure. The conversion and selectivity results with different values of H2 pressure were obtained with the reactor at 673 K, methane pressure at 33.8 kPa and O2 pressure at 8.4 kPa (Foulds and Gray, 1995).

Another factor studied by the authors was the variation in reactor temperature in the presence and absence of H2. The results again showed the system's dependence on the presence of H2, where the highest methane conversion rates were obtained in the presence of H2. Another point observed was that the highest conversion rates were observed at higher temperatures, culminating in the maximum conversion rate at approximately 700°C in the presence of H2 (Foulds and Gray, 1995).

In 1996, LU and colleagues studied the direct oxidation of methane to methanol in a catalytic membrane reactor (CMR) and a fixed bed reactor (FBR). To build the reactors, the authors prepared inorganic microporous membranes made from SiO2/ceramic using the sol-gel method. The results showed that the CMR reactor is more susceptible to temperature increases, as the methane conversion rate increases with increasing reactor temperature and air pressure, reaching maximum values at 700°C and 1.00 kPa of air (Lu *et al*, 1996).

In the general comparison, with the methane conversion rate at 1%, the CMR reactor obtained 11.2% selectivity for methanol, while the FBR reactor obtained 4.5%. When the conversion rate was increased to 3%, the CMR reactor obtained 3% selectivity for methanol and the FBR reactor obtained close to zero selectivity for methanol formation (Lu *et al*, 1996).

The direct oxidation of methane to methanol was also studied by RAJA and RATNASAMY in 1997. The authors studied the conversion of methane to methanol and formaldehyde using O2/tert-butyl hydroperoxide (TBHP) as an oxidant, under ambient conditions, under iron, copper and cobalt phthalocyanines encapsulated in zeolites. For the experiments, a Parr reactor was used at room temperature, where the solid catalysts were added in specific solvents during the experiments (Raja and Ratnasamy, 1997).

During the experiments, various catalyst configurations were tested and the results showed that the type of catalyst used directly interferes with the conversion rate and directs the selectivity in methane oxidation. The best methane conversion results were obtained using the FeCl16Pc catalyst with a conversion rate of 5.3%. With regard to selectivity, the CuCl16Pc-Na-Y(0.11) catalyst obtained 53.5% for the formation of methanol, the CoCli6Pc catalyst obtained 80.5% for the formation of formaldehyde and the Cu(NO2)4Pc catalyst obtained 90% for the formation of ferric acid. On the other hand, several catalysts (K-L, Na-Y, Na-X) obtained values close to zero for methane conversion and in these cases all the methane oxidized was CO_2 (Raja and Ratnasamy, 1997).

When TBHP was used in combination with catalysis, there was a slight improvement in the results, reaching 63.2% selectivity for the formation of methanol when the 5:1 CuCl16Pc:TBHP fraction was used (Raja and Ratnasamy, 1997).

In 1998, AOKI and colleagues studied the direct conversion of methane to methanol on MoO3/SiO2 in two different forms of support. The impregnation form used (NH4)6Mo7O244H2O and silica powder prepared in ethylsilicate hydrolysis and the other form used was the sol/gel method, where (NH4)6Mo7O244H2O was dissolved in ethylene glycol and ethylsilicate, both techniques were calcined at 873 K for 3 hours (Aoki *et al*, 1998).

The results showed that the impregnation system was more effective in converting methane, 40%, while the sol/gel system achieved 12%; but when it came to methanol selectivity, the sol/gel system achieved higher values for methanol formation, 13%, compared to the impregnation system which achieved 2%, in both cases the working temperature was close to the minimum used during all the experiments, 500°C (Aoki *et al*, 1998).

In 2000, GANG and colleagues studied the direct oxidation of methane to methanol using a homogeneous reaction system in oil with $HgSO_4$, relating it to concentration time, increase in gas pressure, catalyst concentration and different pressures. The reactor used was a high-pressure autoclave model with a reaction volume of 200 mL. The overall results showed that the best configuration for methane oxidation was 0.1 % catalyst with a pressure of 31 Bar at 0.74 Bar s^{-1} reaching 97 mmol L^{-1} of methanol (Gang *et al*, 2000).

The authors concluded that the methane oxidation reaction is first order and that the speed is proportional to the concentration of $HgSO_4$, where the reaction limit is given by the solubility limit of the catalyst in the oil (Gang *et al*, 2000).

Also in 2000, PARK and colleagues studied the correlation between catalytic activity and selectivity in methane functionalization. The catalysts used were 5% Pd/C and 5% Pd/AC and the co-catalysts were CuCl2 and $Cu(CH_3COO)_2$ with and without the presence of NaCl. The reactor used was a glass-lined autoclave with a total volume of 300 mL and an operating temperature of 373 K (Park *et al*, 2000).

The results presented showed that the best results were obtained with the PdCl2 catalyst (without support) and CuCl2 as a co-catalyst, reaching 0.49 mmol of formic acid, but when only $PdCl_2$ is used, the selectivity in the formation of formic acid goes up to 0.74 mmol. With regard to the use of NaCl, the best results showed that a [NaCl]/[Pd] ratio of 10.0 is the best composition, reaching 0.79 mmol for the formation of formic acid (Park *et al*, 2000).

In 2002, ZHANG and collaborators studied the controlled partial oxidation of methane on metal oxide catalysts, using a tubular quartz reactor with operating temperatures between 380 - 530°C and 5.0 MPa. As a reference, experiments were carried out with inactive silica in place of the catalysts, with the aim of studying the behavior of the reactor without the catalysts (Zhang *et al*, 2002).

The results show that the quartz tubular reactor has the capacity to oxidize methane without the use of catalysts, achieving 40% selectivity for methanol plus formaldehyde at 460°C (Zhang *et al*, 2002).

When using catalysts, the best results were obtained with the Cat-E catalyst (8.3% Mo + 8.3% V+ 8.3% Cr + 11.1% Bi + 63.9% Si), achieving 70.5% selectivity for methanol plus formaldehyde at 460°C and 4.69% yield for methanol plus formaldehyde at the same temperature (Zhang *et al*, 2002).

Also in 2002, PANTU and GAVALAS studied the partial oxidation of methane in 0.5 % Pt/CeO2 and 0.5 % PT/Al2O3 in a quartz tubular reactor catalyzed at different temperatures (Pantu and Gavalas, 2002).

For the experiments, the authors varied the temperature values and the ratio between methane and oxygen for the two catalysts. The results showed that the higher the ratio between methane and oxygen, the lower the methane conversion, but that the influence of temperature is the opposite, since as the temperature increases, so does the methane conversion (Pantu and Gavalas, 2002).

The results also showed that the 0.5% Pt/CeO2 catalyst obtained better results than the 0.5%

PT/Al2O3 catalyst, achieving 90% methane conversion at 600°C and a working gas of 1.7 (CH4O2), while the 0.5% PT/Al2O3 catalyst obtained 81% under the same experimental conditions (Pantu and Gavalas, 2002).

In 2003, MICHALKIEWICZ used an autoclave reactor to study the conversion of methane into methanol in the condensed phase, using an oil-palladium catalytic system. The reactor was filled with 100 mL of sulfuric acid, 30% SO3 (m/m), 0.3 g of powdered palladium (equivalent to 1.55% m/m) and the methane was pressurized to 4.5 MPa at 160°C. Two types of experiment were carried out: 1) varying the temperature from 90-180°C over 2 hours and 2) varying the experiment time from 1 to 20 hours with a fixed temperature of 160°C (Michalkiewicz, 2003).

The results show that the formation of methanol is dependent on varying the temperature and time of the experiment, where the variation in methanol yield increases gradually with increasing temperature and at higher temperatures there is a rapid increase in yield, reaching maximum values of 20%. When time is varied, there is a rapid increase in the yield of methanol formation in the first few hours and then a stabilization profile is observed in the yield values for the methane to methanol oxidation reaction. Evaluating the results, the authors put the best conditions at temperatures close to 190°C and up to 5 hours of experimentation (Michalkiewicz, 2003).

Also in 2003, WANG and colleagues studied the influence of the V2O5-SiO2 catalyst on the oxidation of methane to oxygenated forms using a fixed-bed flow reactor at different temperatures (Wang *et al*, 2003).

In the first tests, the authors set the working temperature at 625°C and varied the amount of vanadium in the total mass of the catalyst, comparing it with the results for pure silica. The results showed that the best percentage of vanadium is 3%, achieving a selectivity of 8.2% for methanol (181.4 g kg^{-1} of catalyst) and 25.8% for formaldehyde (433.0 g kg^{-1} of catalyst), however, the highest yield for methanol was observed in the experiment with 2% vanadium in the catalyst, 213.0 g of methanol for each kg of catalyst, in these experiments there was a decrease in the yield of formaldehyde, 308.7 g per kg of catalyst (Wang *et al*, 2003).

The authors carried out complementary tests in which the amount of catalyst was set at 2.0% vanadium in V2O5-SiO2 and the reactor temperature was varied. The results showed that the experiments at 550°C achieved the best selectivity results of 19.7% for methanol and 56.0% for formaldehyde, but the best yield results were obtained at 650°C, where the yield values reached 213 g of methanol for each kg of catalyst and 328 g of formaldehyde for each kg of catalyst (Wang *et al*, 2003).

CHEN and colleagues, in 2006, studied the effect of the V2O5 catalyst on the partial oxidation of methane in an autoclave reactor with 250 mL of pure oil and containing 20% and 50% m/m of SO_3, the catalyst concentration varied between 3.5-24.5 mmol of vanadium and the system pressure was 4.0 MPa (Chen *et al*, 2006).

The authors observed that varying the amount of catalyst had a direct effect on the conversion and selectivity values of the process. With regard to conversion, the higher the amount of catalyst the higher the conversion values, reaching 70% at 0.025 mol of catalyst; with regard to selectivity, the performance was different, the higher the concentration of catalyst the lower the selectivity of the system, reaching the highest values at the lowest amount of catalyst, 0.005 mol (Chen *et al*, 2006).

With regard to temperature variation, the same behavior was observed by the authors: increasing the temperature increases the conversion rate, a maximum of 70%, but decreases the selectivity, a minimum of 30% at 485°C.[31]

Another evaluation by the authors is related to the amount of SO3 in the synthesis oil. Increasing this compound in the oil led to an increase in both selectivity and methane conversion, reaching a maximum of 54.5% for methane conversion and 83.5% for system selectivity, both with 30% SO3 in the synthesis oil (Chen *et al*, 2006).

In 2008, two papers were published with the aim of forming methanol from methane via bioprocesses. TABATA and OKURA studied the application of *M. trichosporium* OB3b in a reactor with a recirculation system, the results showed the formation of approximately 3 mmol L^{-1} of methanol in 3 hours of experiment, this yield is equivalent to an efficiency of 2.9% in the conversion to methanol (Tabata and Okura, 2008). RAZUMOVSKY and collaborators studied the application of *Methylosinus sporium* B-2121 in the oxidation of methane to methanol and the results showed that the process is efficient in the formation of methanol, reaching 11.2 mg L^{-1} of methanol in 180 hours of experimentation, equivalent to a specific production of 20 μg of methanol per hour of experimentation (Razumovsky *et al*, 2008).

Also in 2008, FAJARDO and colleagues studied the oxidation of methane to form formaldehyde using silica-supported iron oxide. The experiments were carried out in a quartz tubular reactor at ambient pressure, different temperatures and different catalysts (Fajardo *et at*, 20008).

The authors observed that only pure silica did not produce methanol, but produced formaldehyde at the highest temperatures, reaching 45% selectivity at 750°C. When silica with iron (0.3%) was used, no formaldehyde was formed, but methanol was formed with 91.1% selectivity at 600°C. Using silica with iron (0.5%) maintained the same conditions as the previous catalyst, 92.8% selectivity for methanol, but at 400°C. In a third set of experiments, the authors used 0.5% dissolved iron and the results showed a significant increase in selectivity for formaldehyde, reaching 87.3% at 400°C during the experiment (Fajardo *et at*, 20008).

The authors concluded that the use of silica-iron with a higher percentage of iron leads to selectivity for methanol and at lower temperatures and the use of dissolved iron leads to synthesis for formaldehyde and also at lower temperatures (Fajardo *et at*, 20008).

KHOKHAR and colleagues, in 2009, studied the oxidation of methane by molecular oxygen formed by a binuclear ruthenium complex. The experiments were carried out with 10 atm of methane pressure and 5 atm of O_2 pressure, totaling 15 atm of atmosphere in the reaction system. Under these conditions, methanol was detected as the main product, reaching values of 27 mmol and formaldehyde was detected in small quantities, reaching values of approximately 2.4 mmol. 0.5 mmol of catalyst was used in both experiments. The methanol formation rate was set at 5 x 10^{-4} mol min^{-1} and the formaldehyde formation rate was 0.45 x 10^{-4} mol min^{-1} (Khokhar *et al*, 2009).

The authors also studied the influence of varying the proportion of the gas phase on the selectivity of the compounds. As the CH_4:O_2 fraction increases, so does the selectivity for methanol and the selectivity for formaldehyde decreases, reaching 100% and 0% respectively for methanol and formaldehyde when using 12.5 atm of CH4 and 2.5 atm of O_2 (Khokhar *et al*, 2009).

3 Materials and Methods

The development of the process of electrosynthesis from a gaseous reagent using a gas diffusion electrode was carried out in three stages. The first stage involved the preparation of the metal oxides used in the construction of the gas diffusion electrodes. In this stage, the characterization of the metal oxides and the EDGs was also carried out. In the second stage, the EDGs were tested in preliminary trials using methane gas to confirm the physical and chemical stability of the electrodes. In the third stage, the EDGs were used in the oxidation of ethylene gas.

3.1 Construction of Gas Diffusion Electrodes

The construction of the EDGs was carried out in three stages; in the first stage, the oxides were prepared in the form of a fine powder; in the second stage, the oxide powders were used to prepare the catalytic mass of the EDGs by mixing with the PTFE suspension and catalysts, and in the final stage, the catalytic mass of metal oxides was used to construct the EDGs by pressing and sintering.

3.1.1 Preparation of metal oxides

The metal oxides were prepared following an adaptation of the polymer precursor method described in the literature for depositing titanium/ruthenium oxides on titanium substrates (Forti *et al*, 2001), but the methodology described in the literature had to be altered to allow the metal oxides to be prepared in the form of a fine powder.

For the preparation of the metal oxides, a mass ratio of 70% titanium oxide and 30% ruthenium oxide was set, $(TiO_2)_{0.7}/(RuO_2)_{0.3}$. In order to obtain the oxides in this ratio, a 3:1:1 ratio was established between the reagents ethylene glycol, citric acid and titanium isopropoxide, respectively. The procedure was as follows:

(1) heating ethylene glycol to 65°C under agitation

(2) addition of citric acid under constant stirring

(3) heating the mixture to 90°C while stirring

(4) slow addition of titanium isopropoxide under vigorous stirring until completely dissolved and keep stirring at 90°C for 2 minutes.

(5) addition of ruthenium chloride (alcoholic solution 0.2 mol L^{-1}), stirring and heating for 10 minutes.

At the end of the preparation process, the solution was transferred, while still hot, to a flat-bottomed porcelain plate with a diameter of 15 cm. The porcelain plate was placed in a muffle furnace, EDG model 3000T-S, with a slow heating program for the calcination process of the metal oxides: 1°C per minute up to 50°C for 3 hours, 1°C per minute up to 150°C for 3 hours and 1°C per minute up to 400°C for 1 hour. A flow of 2 mL per minute of pure O_2 was applied from 300°C until the end of the calcination process. After the end of the process, the porcelain plate was kept in the muffle furnace until it cooled to room temperature. The oxides were then ground into a fine powder using a mortar and pestle.

3.1.2 Preparation of Catalyzed Metal Oxides

EDGs were constructed containing titanium/ruthenium oxides catalyzed with different proportions of vanadium, palladium and silver. Proportions of 20%, 10%, 5% and 1% were used for each catalyst. This percentage was calculated in relation to the amount of ruthenium, which together with titanium had fixed proportions.

With regard to the process of preparing the catalyzed metal oxides, the same procedures were followed as described in the previous item, with the addition of the catalysts in step (5) of adding the ruthenium chloride, using the reagents vanadium chloride (solid), palladium chloride solution in HCl (10% mv) and aqueous solution of silver nitrate 1 mol $L^{.-1}$

3.1.3 Preparation of the EDG catalytic mass

With the metal oxides calcined and ground into a fine powder, catalytic masses were prepared with and without the catalysts, following the procedure described in the literature (Forti *et al*, 2007; Beati *et al*, 2009; Rocha *et al*, 2009).

The metal oxides were solubilized in 20 mL of distilled water and stirred for 5 minutes, after which 20% of the hydrophobic binder Dyneon TF 3035 PTFE (Polytetrafluoroethylene) from 3M was added, dispersed at 60%. After adding the PTFE, the solution was stirred for 10 minutes to better homogenize the oxides with the PTFE and then isopropyl alcohol was added under stirring for 10 minutes. At the end of the process, the solution containing the oxides, PTFE and alcohol was transferred to an oven at 120°C for 24 hours to dry the catalytic mass completely.

3.1.4 Construction of Gas Diffusion Electrodes

Once the catalytic mass was completely dry, the EDGs were built using a steel mold (Figures 4 and 5) to obtain electrodes with a diameter of 1 cm. The steel mold comprises (A) the punch, (B) the body of the mold, (C) the EDG holder, (D) the thermocouple hole and (E) the position of the EDG.

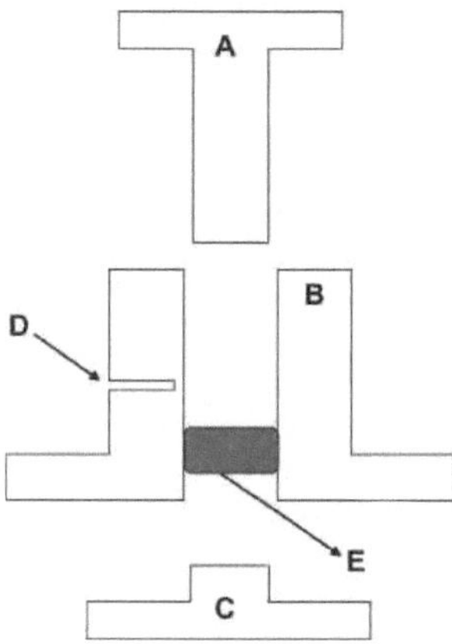

Figure 4 - Steel mold for EDG construction: (A) pulse, (B) mold body, (C) EDG support, (D) thermocouple hole and (E) EDG position.

of the EDG, (D) thermocouple hole and (E) position of the EDG

Figure 5 - Image of the steel mold for building the EDG: (A) pulse, (B) mold body and (C) EDG support

The catalytic mass is placed inside the mold and on top of the EDG support and, with the help of a glass rod, the mass is positioned correctly in a flat and regular manner. The mold is placed inside a refractory oven (Figure 6 A) connected to a temperature controller (Figure 6 B) and both are placed in a hydraulic press, as shown in Figure 6.

Figure 6 - Image of the hydraulic press: (A) furnace and (B) temperature controller

In the EDG sintering process, the steel mould with the catalytic mass is pressed at 200 kgf cm^{-2} for a few seconds and then the pressure is released. The purpose of this process is to shape the electrode without starting the sintering process. Once the pressure has been released, the heating system is switched on and when the mould reaches 120°C, the working pressure (200 kgf cm^{-2}) is applied and the two-hour EDG sintering process begins. The EDG sintering time of 2 hours starts to be recorded from 290°C, but the temperature control system is programmed for a maximum temperature of 340°C. After 2 hours, the heating system is switched off and the EDG remains under pressure until it reaches room temperature.

1.2 Physical Characterization of Metal Oxides and EDG

The physical characterization process was carried out in two stages. In the first stage, the particle size of the oxide powder was determined after the calcination process and before the preparation of the catalytic paste, and in the second stage the morphology and composition of the EDG surface was analyzed. Both stages were carried out using a Jeol JXA840A Electron Probe Microanalyzer scanning electron microscope (SEM), with an energy of 20 KV and magnification of 250x and 500x.

The oxides in powder form were also characterized using a Rigaku model DMAX 2200 X-ray diffractometer to determine which crystallographic structures would be present after the calcination process.

1.3 Determining the flow of methane and ethylene gases in the EDG

In order to carry out the electrochemical oxidation tests using EDG, it was necessary to first quantify the actual amount of gas passing through the electrode during the experiments. To do this, a hollow metal mold (Figures 7 and 8) was used to allow the gas to pass through the electrode, and connected to a column to measure the flow by bubble counting.

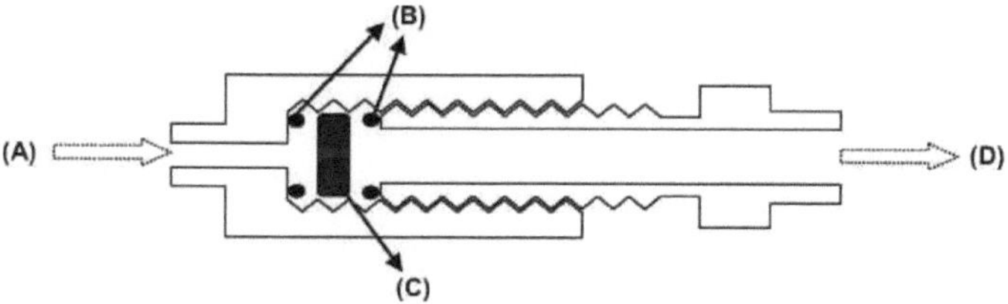

Figura 7 - Schematic representation of the casting of the EDG for determining the permeability: (A) gas inlet, (B) rubber rings, (C) the EDG and (D) gas outlet

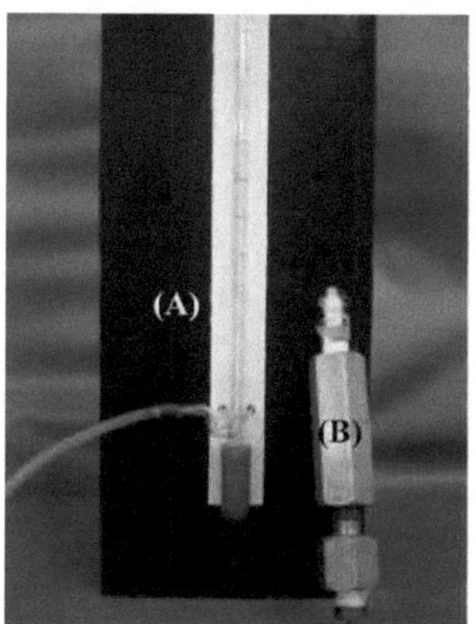

Figura 8 - System for determining permeability: (A) pipette and (B) hollow mold for EDG support

Before the permeability tests, the EDGs went through the complete activation process (item 3.4.1) and, after this process, the EDG was placed in the mold (Figure 8) which was connected to the working gas cylinder (Figure 8 A). The gas was forced to pass only through the EDG structure (Figure

8 C) due to the rubber sealing rings (Figure 8 B) positioned before and after the electrode. At the outlet, the gas flow is recorded. The gas was pressurized at pressures of 0.01 Bar, 0.05 Bar, 0.08 Bar, 0.1 Bar, 0.13 Bar, 0.15 Bar, 0.18 Bar and 0.20 Bar. For each pressure, we waited 20 minutes before taking the flow measurements. This time was necessary for the pressure to stabilize before the flow measurements began.

Figures 9 and 10 show the calibration curves for the variation in volume of methane and ethylene, respectively, as a function of working pressure.

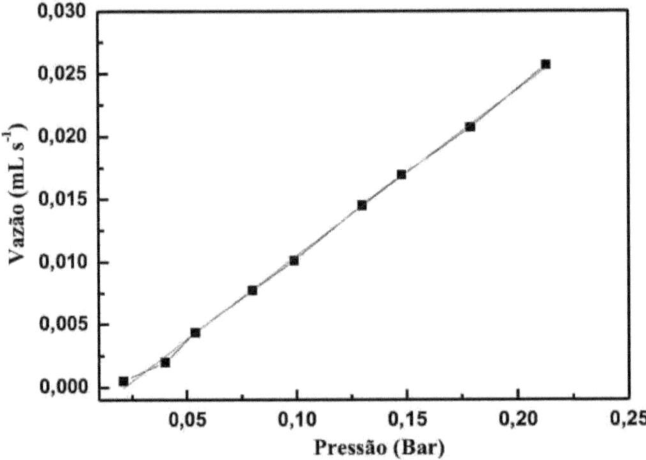

Figura 9 - Variation of gas flow through the EDG as a function of applied pressure

Figure 26 shows the variation in methane flow as a function of the methane gas pressure used (r^2 = 0.99869). To quantify the gas supplied through the EDG, the equation of the straight line of the pressure/flow variation represented by Equation 13 was used:

$$\mathbf{Q = (0{,}13278 \times P) - 0{,}00284} \qquad (13)$$

where,

Q = flow rate $(mL\ s)^{-1}$

P = methane pressure (Bar)

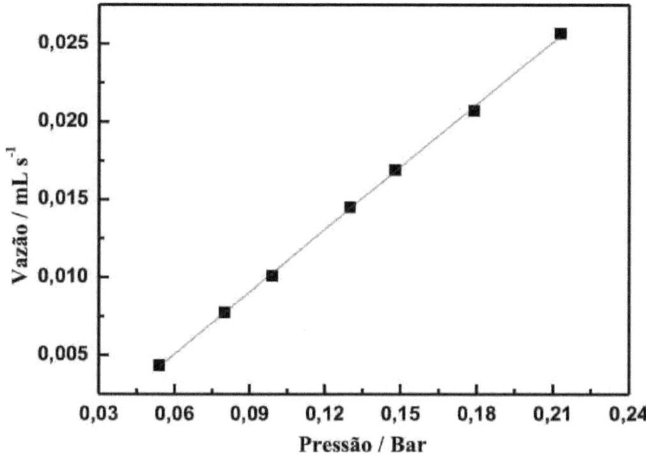

Figura 10 - Variation of ethylene gas flow through the EDG as a function of applied pressure

It can be seen in Figure 10 that the relationship between the flow rate and the gas pressure used is linear (r^2 = 0.99897) throughout the pressure range studied and that the working pressure range (approximately 0.04 Bar) is within this linear range. To quantify the gas supplied through the EDG, the equation of the pressure/flow variation line represented by Equation 14 was used.

$$Q = (0{,}1337 \times P) - 0{,}00297 \quad (14)$$

where,

Q = flow rate (mL s $)^{-1}$

P = applied pressure (Bar)

3.4 Electrochemical tests

3.4.1 EDG activation process

In order to use gas diffusion electrodes, the electrode structure must first be activated. To do this, a one-compartment electrochemical cell with three electrodes was used, with (A) the EDG as the working electrode, (D) the platinum counter electrode and (E) the calomel electrode (ECS) as the reference electrode, as shown in Figure 11. The electrolyte used was 20 mL of Na2SO4 0.1 mol L^{-1} . The electrochemical cell was connected to an Autolab model PGSTAT 20 potentiostat interfaced with a microcomputer.

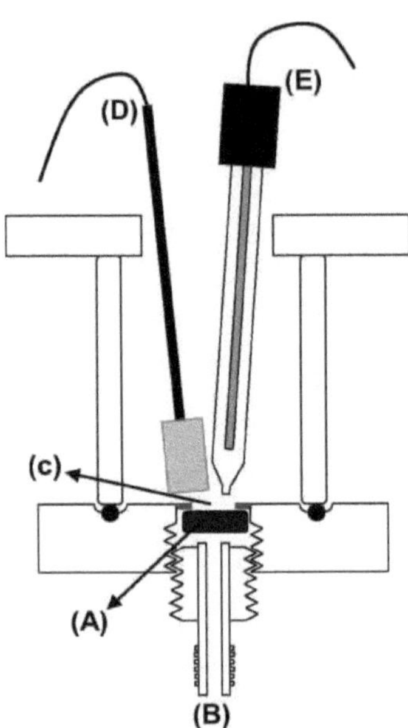

Figure 11 - Schematic representation of the electrochemical cell: (A) EDG, (B) gas chamber, (C) 0.9 cm diameter
electrode opening, (D) platinum counter electrode and (E) ECS reference electrode.

For the activation of the EDG, the Chronoamperometry technique was used, in which a potential of 1.0 V vs. ECS was applied for 1 hour with a constant flow of N2 at a pressure of 0.05 Bar, the activation process was considered complete when there was no variation in the current as a function of the experiment time.

3.4.2 Electrochemical characterization of metal oxide EDGs

The gas diffusion electrodes constructed with $(TiO_2)_{0.7}/(RuO_2)_{0.3}$ were characterized at the oxygen evolution potential using the linear potential sweep. The service life of the EDG was evaluated using chronopotentiometry. The experiments used the electrochemical cell as shown schematically in Figure 8 and without the pressurization of any gas.

The electrochemical technique of cyclic voltammetry was used to evaluate the reversibility of the redox reactions in the EDG, in the potential range of 0.2 V to 1.1 V vs. ECS with a scan rate of 20 $mV.s^{-1}$ and the electrolyte used was H2SO4 0.1 mol L^{-1} . 200 cycles were carried out, where the anodic (Q-) and cathodic (Q+) charges were recorded every 10 cycles, thus evaluating the variability of the charges and also the relationship between the charges (Q+/Q-).

In order to study the O_2 evolution reaction in the EDG, linear sweep voltammetry was used in the potential range from 0 to 3 V vs. ECS, with a sweep speed of 20 $mV.s^{-1}$ and the electrolyte used was H2SO4 0.1 mol L^{-1} . 2 voltammetries were carried out in each experiment, with the second voltammetry being recorded.

In order to evaluate the extent to which the EDG resists with its physical and electrochemical properties unchanged under accelerated working conditions, chronopotentiometry was used, applying a current of 100 $mA.cm^{-2}$ until the measured potential reached 6 V vs. ECS and the electrolyte used was H2SO4 0.5 mol L .$^{-1}$

3.4.3 Electrochemical oxidation of methane

The electrochemical oxidation of methane was carried out on electrodes $(TiO\ )_{20,7}$ /$(RuO\ )_{20,3}$ and electrodes $(TiO\ )_{20,661}$ $(RuO\ )_{20,283}$ $(V2O\ )_{50,056}$, $(TiO\ )_{20,679}$ $(RuO\ )_{20,292}$ $(V2O\ )_{50,029}$ and (TiO2)0.689(RuO2)0.296(V2O5)0.015, corresponding to 0, 20, 10 and 5% vanadium oxide. The experiments used the electrochemical cell shown schematically in Figure 9.

For the study of methane oxidation using EDG of $(TiO\ )_{20,7}$ $(RuO\ )_{20,3}$, the electrochemical technique of chronoamperometry was used in the power range of 1.4 V to 2.2 V vs. ECS, for 1 hour in a support electrolyte of Na_2 SO_4 0.1 mol L^{-1} . For the experiments using EDG of $(TiO\ )_{20,7}$ $(RuO\ )_{20,3}$ catalyzed with vanadium oxide, the electrochemical technique of chronoamperometry was used in the potential range of 1.3 V to 2.3 V vs. ECS for 1 hour in the same support electrolyte. During all the experiments, cell potential and current values were recorded in order to calculate the electrical efficiency and energy consumption of methane oxidation.

For the experiments using $(TiO\ )_{20,7}$ $(RuO\ )_{20,3}$ EDGs, a sample was taken at the end of the experiment at each potential applied, and for the experiments using vanadium-catalyzed EDGs, samples with a volume of 150 μL were taken every 5 minutes for the first 30 minutes of the experiment and every 15 minutes for the final 30 minutes.

3.4.4 Electrochemical oxidation of ethylene

The electrochemical oxidation of ethylene was carried out on electrodes $(TiO\ )_{20,7}$ /$(RuO\ )_{20,3}$ and electrodes catalyzed with 20 % catalyst $(TiO\ )_{20,661}$ $(RuO\ )_{20,283}$ $(MO)_{0,056}$, 10% catalyst $(TiO\ )_{20,679}$ $(RuO\ )_{20,292}$ $(MO)_{0,029}$, 5% catalyst $(TiO\ )_{20,689}$ $(RuO\ )_{20,296}$ $(MO)_{0,015}$ and 1% catalyst $(TiO\ )_{20,698}$ $(RuO\ )_{20,299}$ $(MO)_{0,003}$ where MO stands for AgO, PdO_2 and V O_{25} . The experiments used the electrochemical cell shown schematically in Figure 9.

To study the efficiency of EDG in the oxidation of ethylene, two electrochemical techniques were used: chronoamperometry and chronopotentiometry. For both techniques, 1-hour experiments were carried out and the electrolyte used was Na2SO4 0.1 mol L^{-1} . During all the experiments, the cell potential and current values were recorded in order to calculate the electrical efficiency and energy consumption related to ethylene oxidation.

In the constant potential experiments (chronoamperometry), the potential range of 1.3 V to 2.3 V vs. ECS was used. The constant current experiments used current densities of 0.15 mA cm^{-2} , 0.4 mA cm^{-2} , 0.7 mA cm^{-2} , 1.5 mA cm^{-2} , 5 mA cm^{-2} , 10 mA cm^{-2} and 20 mA cm^{-2} . Samples with a volume of 150 µL were taken every 5 minutes for the first 30 minutes of the experiment and every 15 minutes for the last 30 minutes.

3.4.5 Follow-up Analysis

The samples obtained during electrolysis were inserted into a Varian CP-3800 Gas Chromatograph with a Saturn 2200 Ms/Ms Spectrophotometer, using direct chromatoprobe insertion of 10 µL of each sample. The chromatograph was configured for the detection of methane and ethylene oxidation products, using an unactivated silica column with a length of 2 m and a diameter of 0.1 mm. The injector was heated from 35°C to 250°C at 200°C min^{-1} , the oven was heated from 35°C to 250°C at 100°C min^{-1} , the mass spectrometer was programmed to record masses between 20 and 100 *m/z* with an emission energy of 10 µA.

To quantify the methane and ethylene oxidation products in the GC-MS, methanol, formaldehyde and ferric acid standards were inserted for methane oxidation and ethylene glycol and ethanol were also inserted for the ethylene oxidation reaction. Figures 9 to 11 show the calibration curves for methanol, formaldehyde and ferric acid, respectively.

To construct the methanol calibration curve, 10 standards were inserted at different methanol concentrations, and the calibration curve was constructed based on the count (intensity) of ions at mass 33 *m/z*. The calibration curve for the methanol standard is shown in Figure 12.

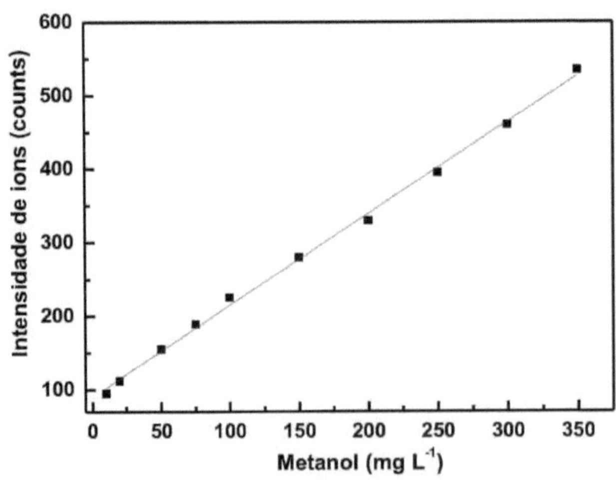

Figura 12 - Calibration curve of methanol standard with monitored ion 33 *m/z*

Figure 12 shows that the relationship between ion intensity and methanol concentration is linear ($r^2 = 0.99799$). To quantify the methanol generated by the oxidation of methane in the EDG, the straight line equation of the variation between ion intensity and methanol concentration was used, represented by Equation 15:

$$C = \frac{ions - 89{,}718}{1{,}448} \tag{15}$$

where,

C = methanol concentration (mg L)$^{-1}$

ions = ion count 33 *m/z* obtained from the mass spectrum of the samples

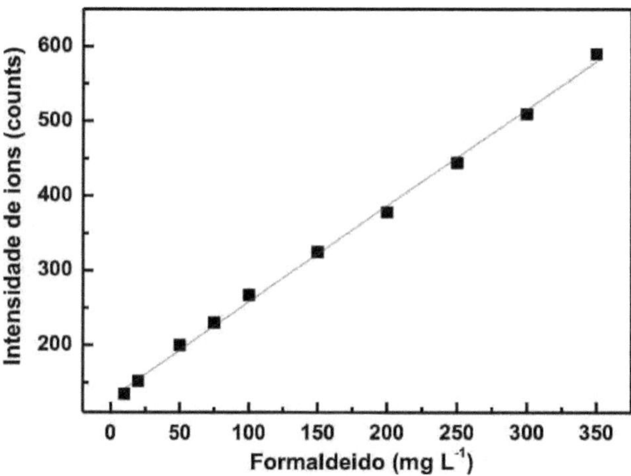

Figura 13 - Calibration curve for the formaldehyde standard

Figure 13 shows the formaldehyde calibration curve. The relationship between ion intensity and formaldehyde concentration is linear ($r^2 = 0.99887$). To quantify the formaldehyde generated by the oxidation of methane in the EDG, the straight line equation of the variation between ion intensity and concentration was used, represented by Equation 16.

$$C = \frac{ions - 129{,}293}{1{,}287} \tag{16}$$

where,

C = concentration (mg L)$^{-1}$

ions = count of ions obtained from the mass spectrum of the samples

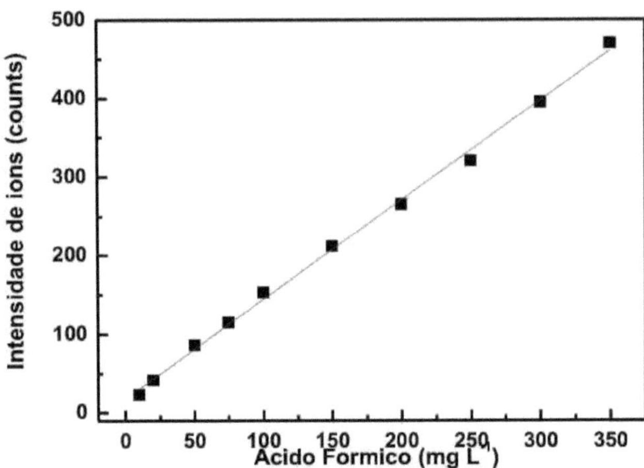

Figura 14 - Calibration curve for the ferric acid standard

Figure 14 shows the calibration curve for ferric acid; the relationship between ion intensity and acid concentration is linear (r^2 = 0.0.99894). To quantify the formic acid generated by the oxidation of methane in the EDG, the straight line equation of the variation between ion intensity and concentration was used, represented by Equation 17.

$$C = \frac{ions - 17{,}781}{1{,}265} \tag{17}$$

where,

C = concentration (mg L)$^{-1}$

ions = count of ions obtained from the mass spectrum of the samples

In the ethylene oxidation experiments, ethylene glycol and ethanol were quantified using a calibration curve with 11 different concentrations of the standards. Figures 15 and 16 show the calibration curve for ethylene glycol and ethanol, respectively.

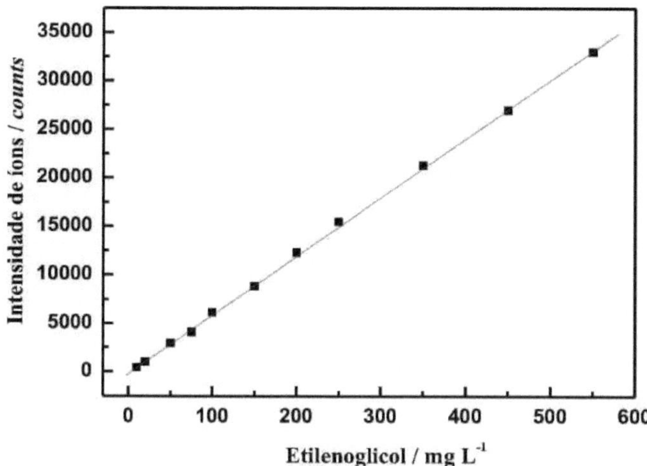

Figura 15 - Calibration curve for ethylene glycol. Range up to 550 mg L^{-1} . ion monitored: ethylene glycol 63 *m/z*

To calculate the concentrations using the calibration curve in Figure 15, the intensity of the 63 m/z ion in the sample was monitored and equation 16 was used to calculate the concentration of ethylene glycol.

$$C = \frac{ions - 30,57}{58,69} \qquad (18)$$

where,

C = ethylene glycol concentration (mg L $)^{-1}$

ions = ion count 63 *m/z* obtained from the mass spectrum of the samples

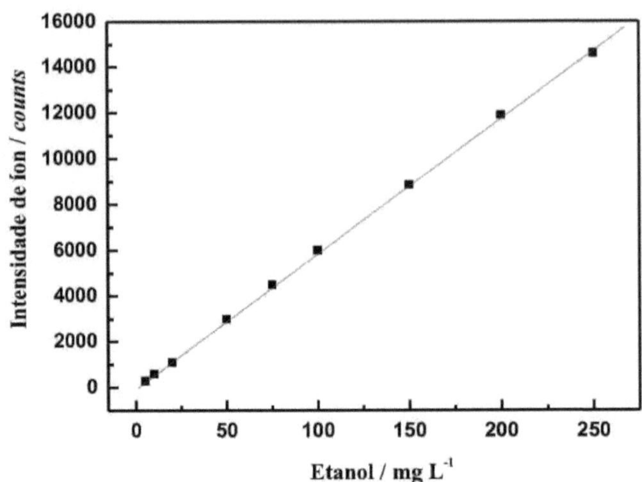

Figura 16 - Calibration curve for ethanol. Range up to 250 mg L^{-1} . ion monitored: ethanol 45 *m/z*

To calculate the concentrations using the calibration curve in Figure 16, the 45 m/z ion intensity of the sample is selected and equation 18 is used to calculate the ethanol concentration.

$$C = \frac{ions - 48,54}{46,75} \qquad (18)$$

where,

C = ethanol concentration (mg L $)^{-1}$

ions = ion count 45 *m/z* obtained from the mass spectrum of the samples

4 Results and Discussion

This chapter presents the results obtained in the three stages of the project: 1) preparation/characterization of the metal oxides and pressing of the oxide EDGs, 2) methane oxidation in preliminary tests of the EDGs and 3) use of the EDGs in ethylene oxidation.

4.1 Physical characterization of metal oxides

The TiO2/RuO2 metallic oxide was analyzed under a scanning electron microscope to determine the particle size of the powder. The micrograph of the TiO2/RuO2 sample is shown in Figure 17.

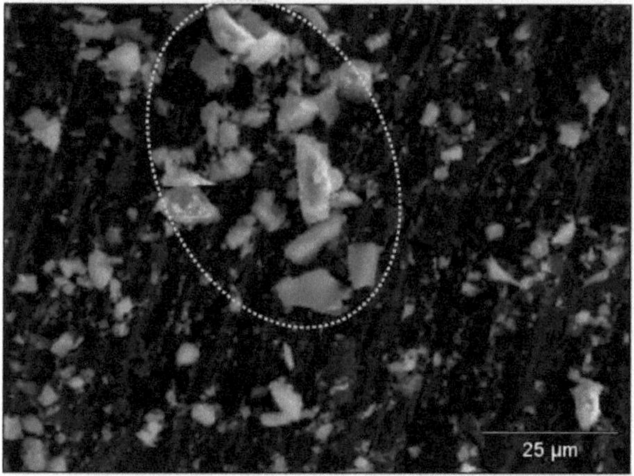

Figure 17 - Micrographic image of TiO2/RuO2 powder. 1000x magnification

Figure 17 shows the micrograph of the TiO2/RuO2 powder deposited on a carbon adhesive tape. There is a predominance of particles between 1 and 4 µm in size, but there are also several larger particles (highlighted in Figure 17). These larger particles may be related to flaws in the grinding of the oxide powder before the calcination process. The size distribution of the TiO2/RuO2 particles can be seen in Figure 18.

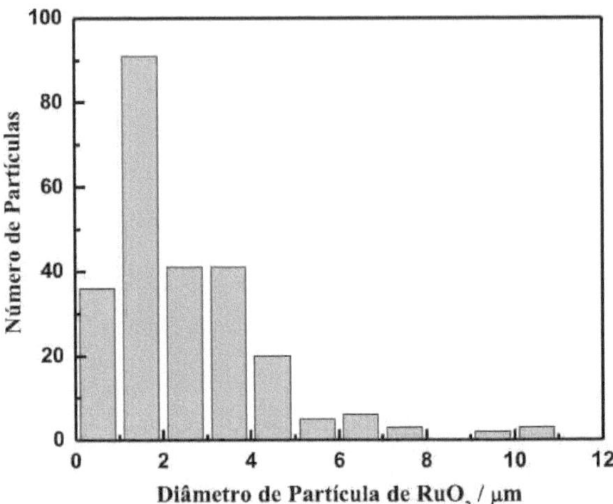

Figure 18 - Size distribution of TiO_2/RuO_2 particles

Figure 18 shows the diameter distribution of the particles shown in Figure 17. It can be seen that the vast majority of the particles have a diameter of between 1 and 4 µm and others between 9 and 11 µm, which are represented in the highlight of Figure 17. The assessment of the particle size of the metal oxide powder was only shown for TiO_2/RuO_2, but the other catalyzed oxides showed the same particle size distribution, with most of the particles remaining in the 1 to 4 µm range.

Looking at Figures 17 and 18, it can be seen that most of the particles are between 1 and 4 µm in size. This size distribution was also observed in all the catalyzed oxides, but there is a need to characterize the composition of the oxides as well as the size. Figures 19 to 22 show the X-ray diffractograms of the TiO2/RuO2 samples and the samples catalyzed with V_2O_5, PdO2 and AgO.

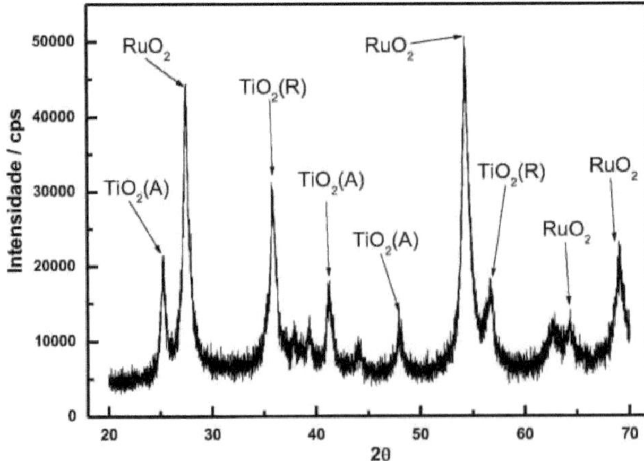

Figure 19 - X-ray diffractogram of the TiO_2/RuO_2 sample

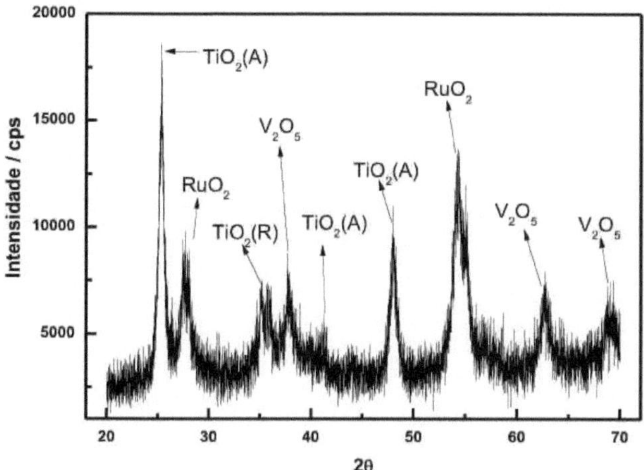

Figura 20 - X-ray diffractogram of the $(TiO_2)_{0.66i}(RuO_2)_{0.283}(V_2O_5)_{0.056}$ (20% Vanadium) sample

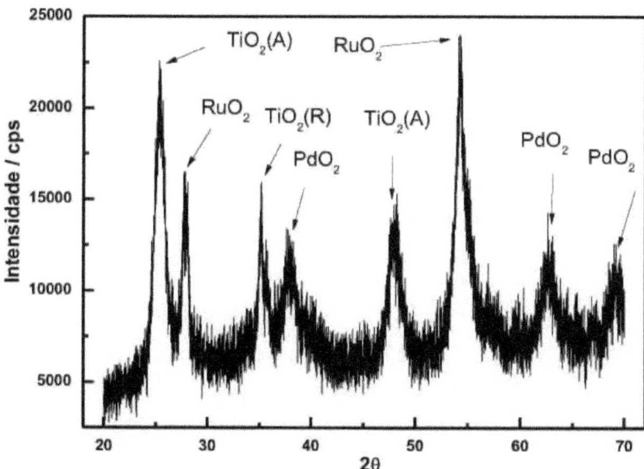

Figura 21 - X-ray diffractogram of the $(TiO_2)_{0.66i}(RuO_2)_{0.283}(PdO_2)_{0.056}$ sample (20% palladium)

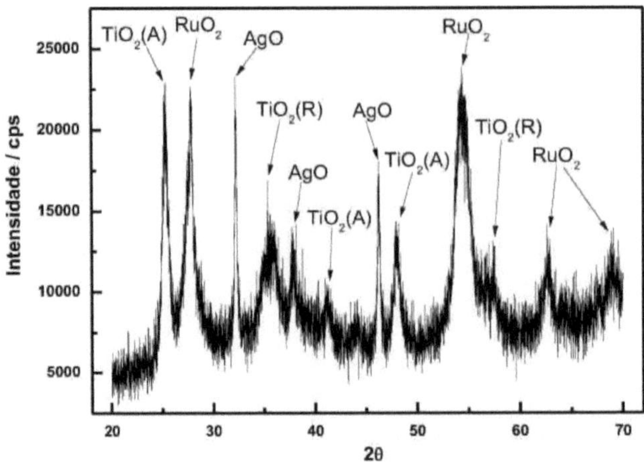

Figura 22 - X-ray diffractogram of the $(TiO_2)_{0.661}(RuO_2)_{0.283}(AgO)_{0.056}$ (20% Silver) sample

In order to apply metal oxide EDGs to the electrochemical oxidation of gaseous organic compounds, it is necessary to first determine the composition of the metal oxides used in the construction of the electrodes.

Figures 19 to 22 show the diffractograms of the metal oxide samples. The 2θ angles for titanium oxide, the anatase phase (A) and the rutile phase (R) can be seen in all the figures. In addition, the presence of peaks relating to ruthenium oxide indicates that the calcination process was successful in forming TiO_2/RuO_2. With regard to the addition of the catalysts, it can be seen that the calcination process achieved its objective of forming vanadium (Figure 20), palladium (Figure 21) and silver (Figure 22) oxides.

Figure 20 shows the 2θ angles for titanium oxide (A and R) and ruthenium oxide. There are also three peaks (37.9; 62.4; 68.7) for vanadium oxide (V_2O_5). Figure 21 shows the same peaks for the titanium/ruthenium oxides as well as three peaks (37.2; 63.5; 69.7) for palladium oxide (PdO_2). Figure 22 shows the 2θ angles for titanium/ruthenium oxides and three angles (32.2; 38.4; 46.1) for silver oxide (AgO).

4.2 Physical Characterization of Gas Diffusion Electrodes

The gas diffusion electrodes made of TiO2/RuO2 were analyzed using a scanning electron microscope with an EDX probe. The non-destructive analysis allowed for the characterization and quantification of the species present on the electrode surface. The micrographs are shown in Figures 23 to 26.

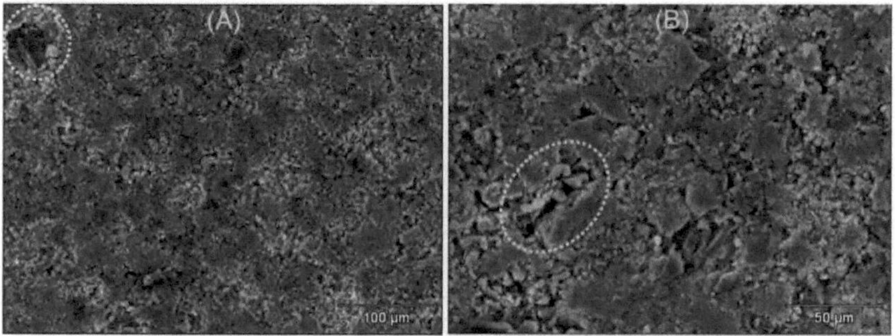

Figure 23 - Micrographic image of TiO_2/RuO_2 EDG at (A) 250x and (B) 500x magnification

Figure 23 shows the surface of the TiO_2/RuO_2 EDG at (A) 250x and (B) 500x magnification. It shows the compact surface of the EDG, free of cracks or large fissures (Figure 23 A), but there are some areas with larger oxide particles (detail of Figure 23 A). These larger particles may impose flaws in the interaction between oxides and PTFE, causing possible preferential passage of gas through the EDG structure, as can be seen in the detail of the enlarged image in Figure 23 B.

Looking at Figures 23 A and 23 B, we can see a big difference between the TiO_2/RuO_2 EDG and conventional TiO_2/RuO_2 flat electrodes. The EDG does not have a surface with cracks similar to cracked clay, and this type of surface is described in the literature for flat electrodes with TiO_2/RuO_2 coating on titanium (Pelegrino *et al*, 2002). It is possible that the EDG does not have a surface similar to cracked clay due to the process of calcining and grinding the oxides before preparing the catalytic mass and constructing the gas diffusion electrode. Similar to the TiO_2/RuO_2 EDG, the surfaces of the electrodes with the addition of the catalysts were also analyzed in order to determine possible changes in the electrode surface with the addition of the catalysts.

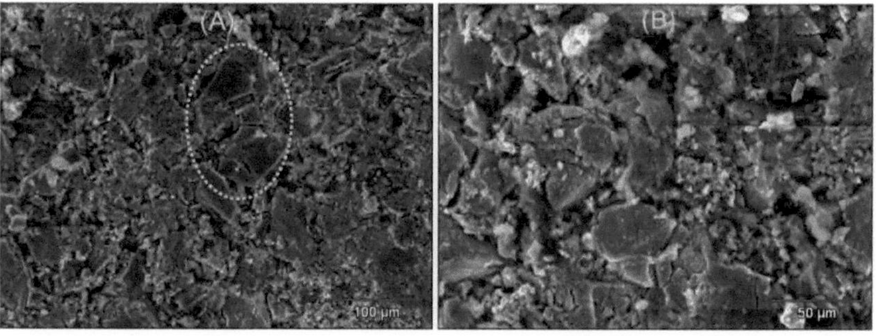

Figure 24 - Micrographic image of the EDG of $(TiO_2)_{0.661}(RuO_2)_{0.283}(V_2O_5)_{0.056}$ (20% Vanadium)

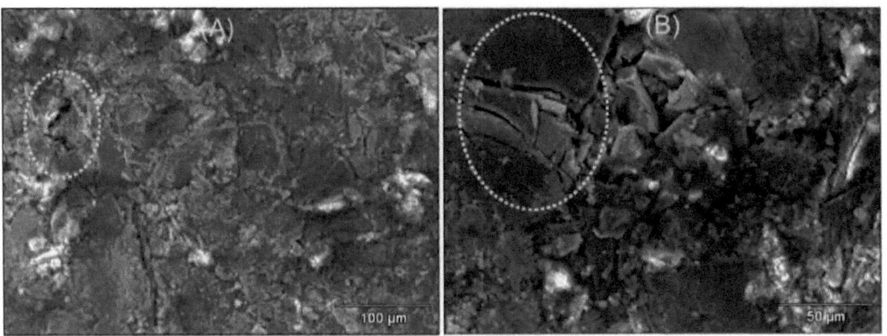

Figure 25 - Micrographic image of the EDG of $(TiO_2)_{0.66i}(RuO_2)_{0.283}(PdO_2)_{0.056}$ (20% palladium)

Figure 26 - Micrographic image of the EDG of $(TiO_2)_{0.661}(RuO_2)_{0.283}(AgO)_{0.056}$ (20% Silver)

Figure 24 shows the influence on the electrode surface, with the addition of vanadium showing a less compact surface compared to the TiO2/RuO2 EDG with the presence of some larger particles (highlighted in Figure 24 A) and the electrode surface also showing holes and spaces between the oxide particles (Figure 24 B), this could lead to flaws in the interaction between the oxides and PTFE, causing possible preferential passage of gases through the EDG structure.

Figure 25 shows micrographs of the titanium/ruthenium oxide EDGs with the addition of the palladium catalyst. These electrodes have a compact surface, like the TiO_2RuO_2 EDG, and also have some larger particles (highlights in Figures 25 A and 17 B). Figure 26 shows the micrographic images of the EDG with silver catalyst, with a compact surface and no cracks or fissures. The presence of a few larger particles can also be seen, as can be seen in the details in Figures 26 A and 26 B.

Observing the surface of the oxide EDGs (Figures 23 to 26), one can generally see compact and uniform structures. However, it is necessary to know the quantities of each of the oxides present on the surface of the electrode and to this end, the EDGs were analyzed by X-ray scattering (EDX) using a scanning electron microscope. The results are shown in Tables 1 to 4.

Table 1 - Theoretical and Real Quantities (%) of titanium and ruthenium oxides

	Titanium		Ruthenium	
	Theoretical	Real	Theoretical	Real
EDG	70	70,4	30	29,6

Table 2 - Theoretical and Real Quantities (%) of titanium, ruthenium and vanadium oxides

	Titanium		Ruthenium		Vanadium	
V2O5	Theoretical	Real	Theoretical	Real	Theoretical	Real
20%	66,1	65,8	28,3	28,5	5,6	5,7
10%	67,9	68,3	29,2	29,1	2,9	2,6
5%	68,9	69,2	29,6	29,2	1,5	1,6
1%	69,8	69,4	29,9	30,1	0,3	0,5

Table 3 - Theoretical and Real Quantities (%) of titanium, ruthenium and palladium oxides

	Titanium		Ruthenium		Palladium	
PdO2	Theoretical	Real	Theoretical	Real	Theoretical	Real
20%	66,1	66,5	28,3	28,0	5,6	5,5
10%	67,9	67,6	29,2	29,5	2,9	2,9
05%	68,9	68,9	29,6	29,8	1,5	1,3
01%	69,8	69,4	29,9	30,2	0,3	0,4

Table 4 - Theoretical and Real Quantities (%) of titanium, ruthenium and silver oxides

	Titanium		Ruthenium		Silver	
AgO	Theoretical	Real	Theoretical	Real	Theoretical	Real
20%	66,1	65,8	28,3	28,4	5,6	5,8
10%	67,9	68,0	29,2	29,3	2,9	2,7
05%	68,9	69,1	29,6	29,4	1,5	1,5
01%	69,8	70,1	29,9	29,7	0,3	0,2

Table 1 to 4 shows the quantities, in %, of the oxides on the EDG surface. It can be seen that the expected theoretical quantities of the oxides are close to the actual quantities observed in the EDX results, but in some catalysts there were significant differences between the theoretical and actual quantities. In the EDG with 1% vanadium, a difference of approximately 66% was observed between the expected theoretical amount (0.3%) and the actual amount of vanadium in the EDX (0.5%). In the EDG with 5% palladium, a difference of 13% in palladium was observed between the theoretical amount (1.5%) and the actual amount (1.3%) and in the EDG with 1% palladium, the difference was approximately 33% between the actual amount 0.4% and 0.3% of the expected theoretical amount of palladium. In the EDG catalyzed with 1% silver, a variation of approximately 33% was observed between the theoretical amount of silver (0.3%) and the actual amount (0.2%).

Having defined the amount of titanium oxide, ruthenium oxide and the oxides of the catalysts on the gas diffusion electrode, we analyzed how these oxides are distributed on the surface of the EDG. The EDX (Energy Dispersive Spectroscopy) mapping technique was used in the scanning electron microscope and the results are shown in Figure 20.

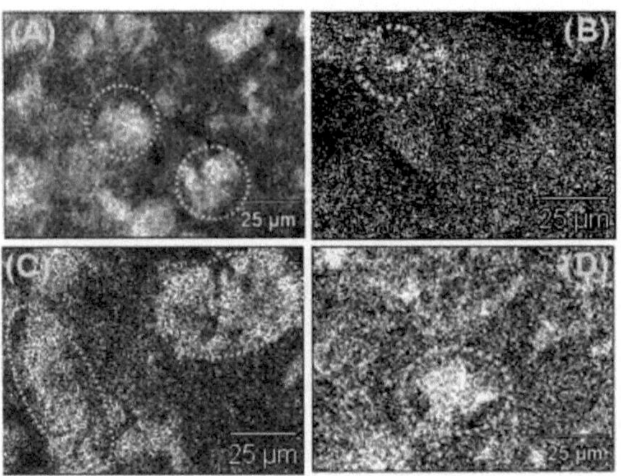

Figure 27 - EDG micrographic image (500x) with EDX mapping, highlighting the light pigments: (A) ruthenium, (B) vanadium, (C) palladium and (D) silver.

Figure 27 shows micrographs of the catalyzed EDGs, where the metal oxides are represented by the light pigments. It can be seen that the distribution of ruthenium (Figure 27 A) over the surface of the EDG is not regular, as there are areas with greater agglomeration of ruthenium, according to the highlights in Figure 27 A. Figure 27 B shows the distribution of vanadium on the EDG surface, which is better distributed compared to the distribution of ruthenium. The micrograph shows only one area (highlighted in Figure 27 B) with a higher concentration of vanadium, but the other areas of the electrode surface show a more regular distribution.

Figure 27 C shows the micrograph of the surface of the palladium-catalyzed EDG with two regions with a higher concentration of palladium oxide, as highlighted in Figure 27 C. Figure 27 D shows the surface of the silver-catalyzed EDG, the electrode with the best distribution of the catalyst metal.

Observing the results of the characterization of the EDGs catalyzed with metal oxides, it can be seen that the electrodes have a compact surface with no large cracks. Another characteristic of the catalyzed EDGs was the concentration of metal oxides on the surface of the particles, which reached values close to the theoretical values expected for each of the catalysts. This showed that the process of homogenization, calcination and preparation of the catalytic mass was effective. Finally, the distribution of the oxides on the EDG surface was analyzed. The results showed that the Ti/Ru electrodes and the electrodes catalyzed with palladium had areas with greater amounts of Ru and Pd, respectively, but the EDGs catalyzed with vanadium and silver had better distributions of these metals, and no large agglomerations of the catalyst metals were observed.

4.3 Electrochemical Characterization of Gas Diffusion Electrodes

In the electrochemical characterization, only the results for the TiO2RuO2 electrodes were presented, since the EDGs with the vanadium, palladium and silver catalysts obtained the same results, and it was not possible to distinguish any changes in the results. The results of the electrochemical characterization of the TiO2RuO2 EDG are presented below.

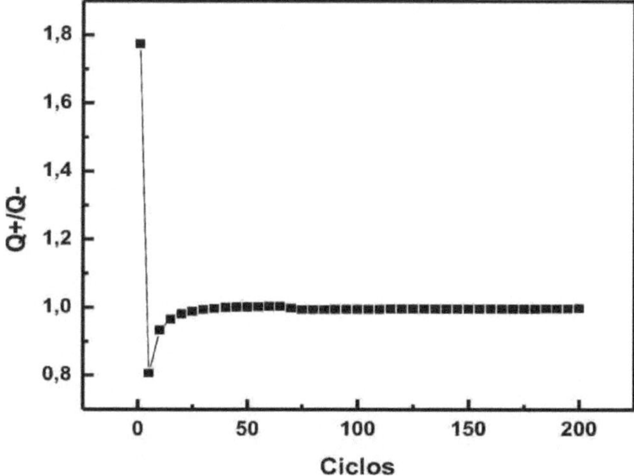

Figure 28 - Variation of the Q+/Q- ratio during 200 cycles. Fixed potential 0.2 V to 1.1 V vs ECS with a scan rate of 20 $mV.s^{-1}$. Electrolyte: 20 mL of H2SO4 0.1 mol L^{-1}

Figure 28 shows the variation in anodic and cathodic charges during 200 cycles of cyclic voltammetry. There is a large variation in the anodic and cathodic charge values up to 25 cycles and a subsequent stabilization. From then on, the Q+/Q- ratio approaches 1. This variation can also be seen in Figure 23, which shows the variation in the Q+/Q- ratio over 200 cycles. There is a large variation in the ratio between the charges up to 10 cycles, after which the charge ratio stabilizes until the end of the experiment, where the variation in the charge ratio remained close to 1.02 until the end of the experiment. This ratio between the anodic and cathodic charges close to 1.02 may be associated with the reversibility of the ruthenium redox reactions on the surface of the gas diffusion electrode.

Figure 29 shows the linear voltammetry in the EDG of oxides with 20% PTFE, the voltammogram started at 0.0 V vs. ECS and moved towards more positive potentials up to 3.0 V vs. ECS.

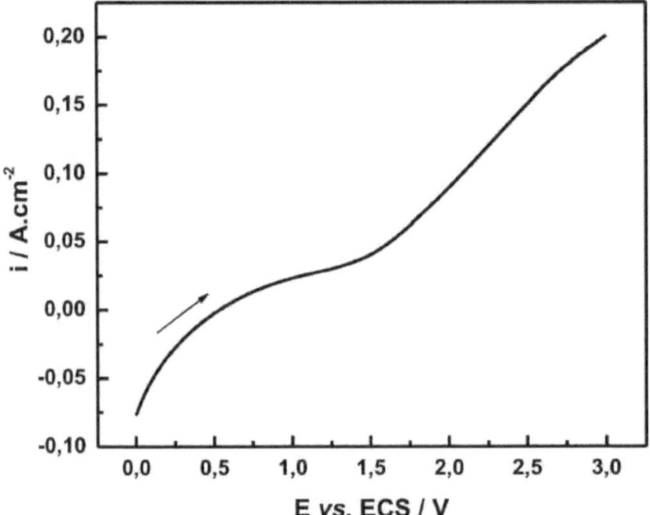

Figure 29 - Linear voltammetry of the EDG of oxides with 20% PTFE, in the potential range 0.0 V to 3.0 V vs ECS with a scan rate of 20 $mV.s^{-1}$. Electrolyte: 20 mL of H2SO4 0.1 mol L^{-1}

There is an increase in current of approximately 1.5 V vs. ECS. This increase is associated with the O_2 release reaction in the EDG. This information is important because the range where O_2 begins to evolve is the region where RuO_3 is formed, the active species responsible for oxidizing the compounds in the oxide EDG (Comninellis, 1997).

As described, the oxide EDG has good symmetry and stability between the anodic and cathodic charges (Figure 28), and also has the property of forming O_2 at approximately 1.5 V vs. ECS. To complete the characterization of the electrode, a lifetime assessment was carried out under working conditions, as shown in Figure 30.

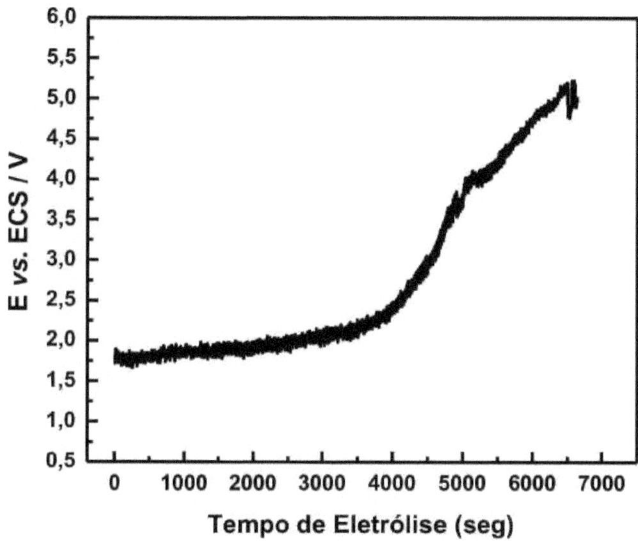

Figure 30 - Chronopotentiometry of the EDG of oxides with 20% PTFE, with an applied current density of
100 mA cm^{-2} Electrolyte: 20 mL of H2SO4 0.5 mol L^{-1}

Figure 30 shows the evaluation of the lifetime under working conditions, at 100 mA cm^{-2} , where the potential remained stable at approximately 2 V vs. ECS for 70 minutes, followed by a rapid increase in potential, reaching approximately 5.5 V vs. ECS in less than 100 minutes. This increase in potential may be associated with the increase in electrode resistivity during the experiment due to the loss of ruthenium and the consequent increase in the proportion of TiO2.

In view of the results presented, there is an excellent stability of the anodic and cathodic charges observed in the relationship between the charges (Figure 27). It can also be seen that the potential for the O2 release reaction is approximately 1.5 V vs. ECS (Figure 29). In the service life test, the electrode maintained a potential of approximately 2 V vs. ECS for 70 minutes (Figure 30). Given these results, the gas diffusion electrode built with the metal oxide powder was ready for the preliminary electrochemical oxidation tests of methane gas.

4.4 Electrochemical Oxidation of Methane Gas

For the initial electrochemical methane gas oxidation tests, the TiO2RuO2 EDG was polarized in the potential range of 1.4 V to 2.2 V vs. ECS. The beginning of the potential range for the experiment was determined by observing the current profile of the O2 release reaction (Figure 29) where the current increases at approximately 1.5 V vs. ECS. In order to determine the end of the potential range for the experiments, the process efficiency results were observed, the reduction in efficiency compared to the efficiency of the previous potential. For the experiments using methane, a pressure of approximately 0.08 Bar was determined, a value that prevents the formation of bubbles on the electrode surface. Excessive accumulation of bubbles on the electrode surface could cause a reduction in the active surface of the EDG. To analyze the electrolyzed samples, a standard sample of methanol, the main expected product, was inserted into the GC-MS and the mass spectrum is shown in Figure 31.

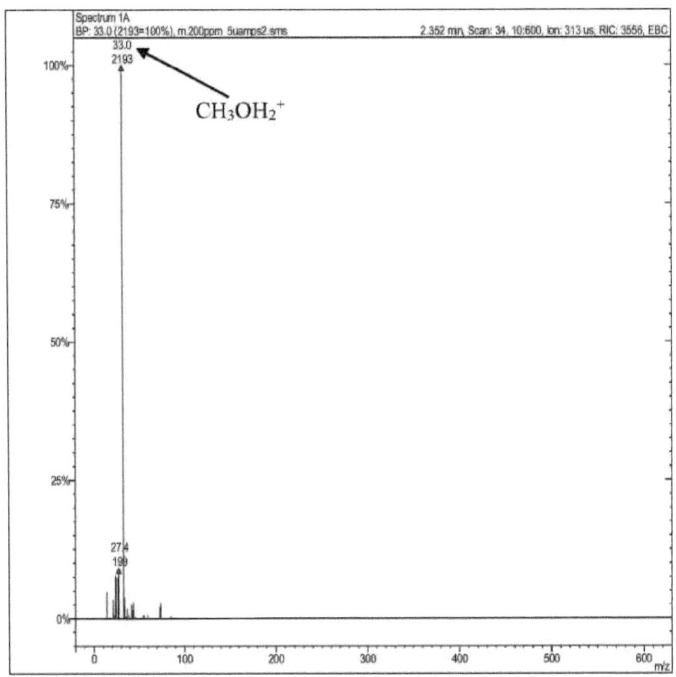

Figure 231 - Mass spectrum of the standard methanol sample

Figure 31 shows the mass spectrum of the standard methanol sample, with the presence of the 33 *m/z* mass referring to the protonated methanol molecule, a phenomenon characteristic of some molecules in the mass spectrometer *iontrap*.

During the experiments at constant potential, the concentration of methanol, resulting from the oxidation of methane, increased with the potential applied, as can be seen in Figure 32. At a potential of 2.2V vs. ECS, the concentration reached 148 mg L $.^{-1}$

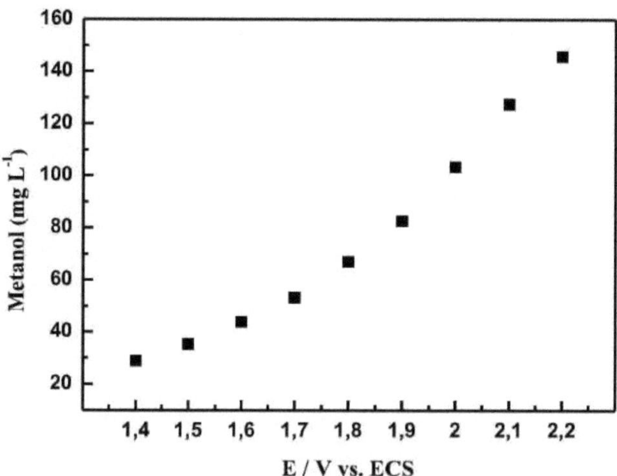

Figura 32 - Variation of methanol concentration as a function of applied potential.
Potential range
1.4 V to 2.2 V vs. ECS. Electrolyte: 20 mL Na2SO4 0.1 mol L^{-1}

The samples were then inserted into a gas chromatograph coupled to a mass spectrometer, and Figure 28 shows the mass spectrum of the sample obtained in the experiment at 2.1 V vs. ECS.

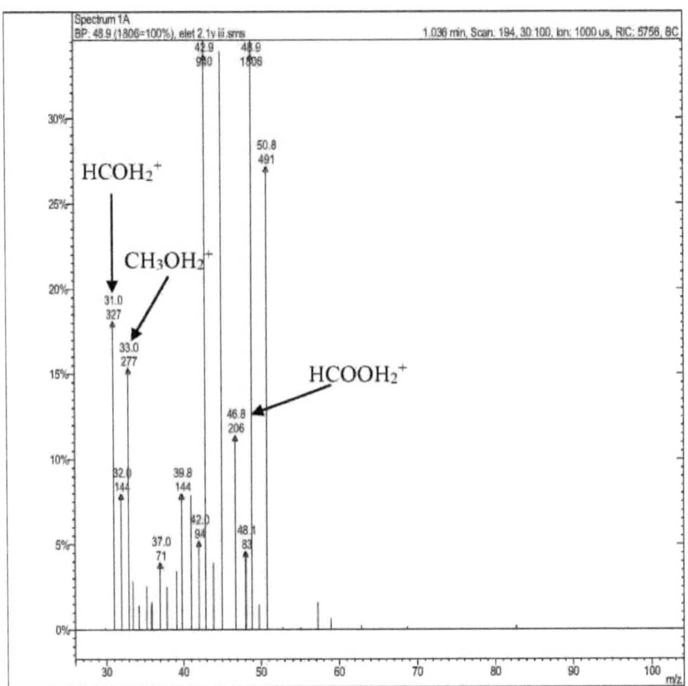

Figura 33 - Mass spectrum of the final sample of the experiment at 2.1 V vs. ECS

Looking at Figure 33, we can see certain masses possibly originating from the methane oxidation reaction in the oxide EDG: 31 *m/z*, referring to protonated formaldehyde; 33 *m/z,* referring to protonated methanol; 46.8 *m/z,* referring to protonated ferric acid. There are also other masses in the mass spectrum, probably fragments of ionization in the mass spectrometer filament. With regard to the mass values expressed in the mass spectrum, some values are not exact, such as 46.8 *m/z* for protonated ferric acid. This variation is expected and determined in the calibration of the GC-MS, which is why, for the purposes of analyzing the results, the mass of the protonated sorbic acid considered and calibrated was 47 *m/z*.

Once the methanol resulting from methane oxidation has been quantified, it is necessary to determine the chemical efficiency of the process. For this calculation, it was necessary to determine the amount of methane gas supplied to the reaction through the electrode structure, using equation 13. With the volume defined, the general law of gases under normal conditions of temperature and pressure (CNTP: T = 20 °C and P = 1 atm) was used to calculate the number of moles equivalent to the volume of gas calculated.

With the number of moles of methane defined, it was compared with the number of moles of methanol formed, and this comparison determined the rate of methanol formed from the methane supplied, the results of which are shown in Figure 34.

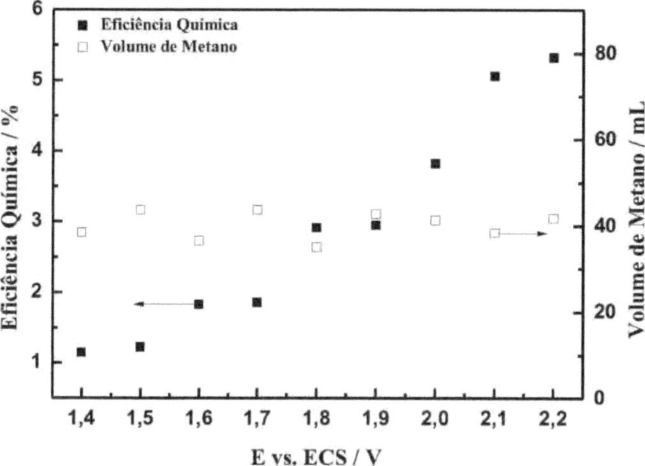

Figura 34 - Variation in chemical efficiency (%) and volume of methane supplied (mL) both in relation to the applied potential

Figure 34 shows the variation in chemical efficiency values as a function of the potential applied, with relatively low values, reaching a maximum of 5.5% for the potential of 2.2 V vs. ECS. These low chemical efficiency values may be associated with the excess gas supplied to the EDG during the experiment.

It can also be seen in Figure 34 that the increase in chemical efficiency is associated with an increase in the applied potential. Another point to be observed in the methane gas oxidation experiments is the efficiency of the electric charge transfer for the methane oxidation reaction to form methanol. Considering that this reaction transfers 2 (Equation 20) electrons, the electric charge needed to form the appropriate concentrations of methanol was calculated (Figure 32).

$$CH_4 + 2H_2O \rightarrow CH_3OH + 4H^+ + 4e^- + 1/2O_2 \qquad (20)$$

The specific electrical charge of methanol was compared with the total electrical charge values of the experiment (data provided by the GPES potentiostat management program) and the results are shown in Figure 35.

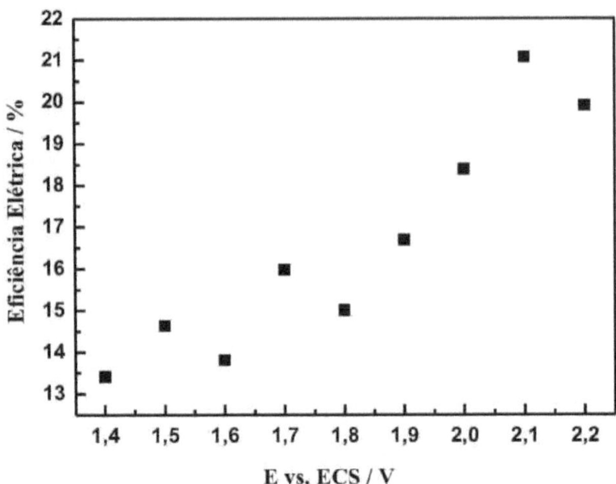

Figura 35 - Variation in Electrical Efficiency (%) as a function of applied potential

Figure 35 shows the variation in electrical efficiency as a function of the applied potential. It can be seen that the efficiency increases with the increase in the applied potential up to 2.1 V vs. ECS; at the higher potential, 2.2 V vs. ECS, there was a slight decrease in electrical efficiency.

With regard to the electrical energy consumption involved in the methane to methanol oxidation reaction, Figure 36 shows the energy consumption as a function of the potential applied.

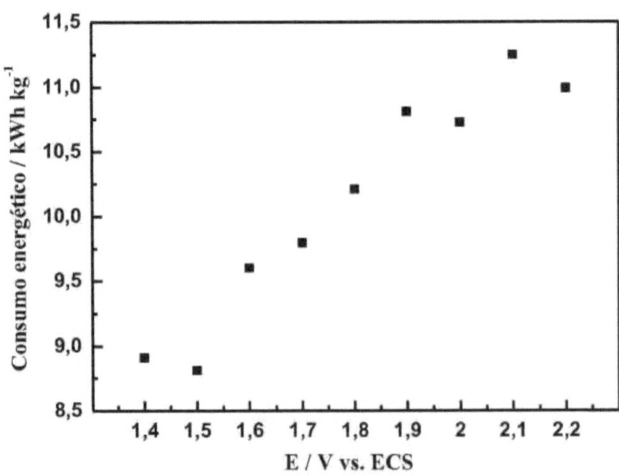

Figura 36 - Variation in energy consumption as a function of applied potential

Figure 36 shows the variation in energy consumption as a function of the potential applied in the experiments to oxidize methane to methanol. It can be seen that energy consumption generally increases as the applied potential increases, reaching 11.2 kWh kg^{-1} of methanol formed. With the results presented, it can be seen that the gas diffusion electrode built with titanium oxide and ruthenium is efficient in oxidizing methane gas to form methanol and other by-products of the reaction, which is why a new constant current experiment with oxide EDG was planned for methane oxidation and the results are shown in Figure 37.

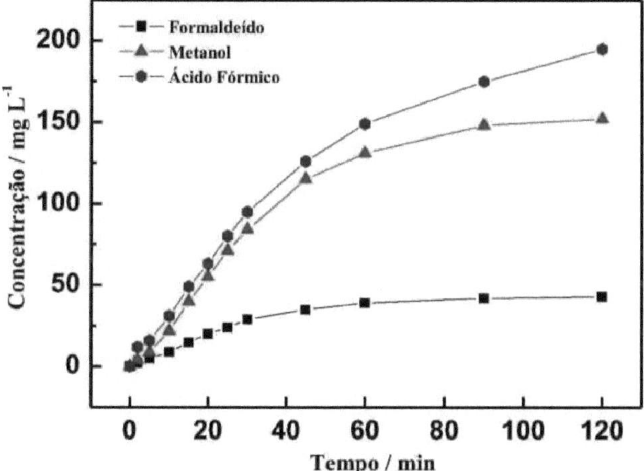

Figura 37 - Variation in the concentration of methanol, formaldehyde and thermal acid as a function of the

current density of 8.5 mA cm^{-2} . Electrolyte: 20 mL Na2SO4 0.1 mol L^{-1}

Figure 37 shows the variation in the concentration of methanol, formaldehyde and formic acid as a function of the current density applied. It shows the formation of 43 mg L^{-1} of formaldehyde, 152 mg L^{-1} of methanol and 195 mg L^{-1} of formic acid at the end of the two-hour experiment at 8.5 mA cm^{-2} . It can be seen in Figure 32 that the constant current experiment formed a higher concentration of ferric acid compared to methanol and formaldehyde. This higher acid formation may be associated with the current density applied in the experiment, promoting higher acid formation and lower aldehyde formation.

With the results presented, the TiO2/RuO2 EDG is efficient in the oxidation of methane gas, forming methanol, formaldehyde and ferric acid. Due to the results obtained, methane oxidation experiments were carried out using TiO2/RuO2 EDG catalyzed with different amounts of vanadium oxide.

4.4.3 Methane oxidation using TiO2/RuO2/V2O5 EDGs

In view of the results obtained, it was decided to add vanadium (in oxide form) to the TiO2RuO2 EDG in various concentrations as a strategy to study the influence of the catalyst on the formation of by-products and the efficiency of methane oxidation.

In the experiments with vanadium-catalyzed EDG, a pressure of approximately 0.04 Bar was used in order to provide a smaller quantity of gas, close to the stoichiometric quantity of the methane oxidation reaction.

Another point evaluated was the range of potential applied: in the experiments with TiO2RuO2 EDG, the range 1.4 V to 2.2 V vs. ECS was used and it was observed that the methanol concentration increased with the increase in potential applied, as did the current efficiency. In order to observe the evolution of current efficiency over a wider potential range, the 1.3 V to 2.3 V vs. ECS range was used.

- EDG with 20% vanadium oxide

In the methane oxidation experiments using the EDG of $(TiO_2)_{0.661}(RuO_2)_{0.283}(V_2O_5)_{0.056}$, it was decided to quantify all the products identified in the methane oxidation process.

Figure 38 shows the variation in methanol concentration as a function of the potential applied during a one-hour experiment, where it can be seen that as the potential applied increases, methanol formation increases. At a potential of 1.3 V vs. ECS, 150 mg L^{-1} was formed and at a potential of 2.3 V vs. ECS, 340 mg L^{-1} was formed.

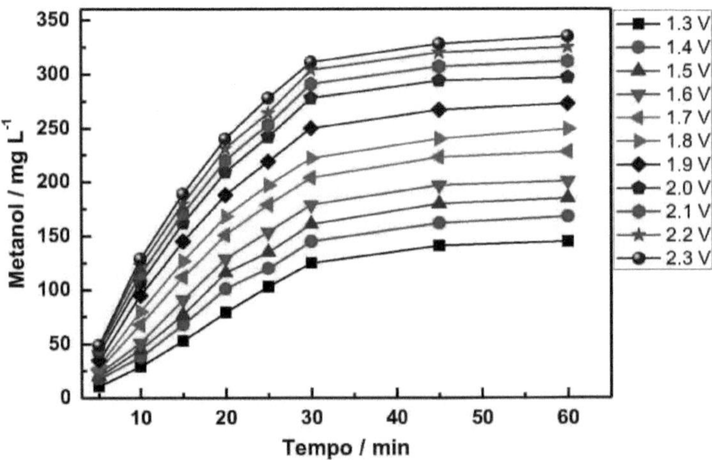

Figura 38 - Variation of methanol concentration as a function of experiment time using EDG with 20% vanadium. Electrolyte: 20 mL Na2SO4 0.1 mol L^{-1}

Figure 38 also shows the variation in methanol concentration during the experiments, where you can see the stabilization profile of the variation in methanol concentration as a function of time. The concentration in all experiments increases rapidly up to 30 minutes, followed by a stabilization of concentrations in the last 30 minutes of the experiment. This stabilization profile may be associated with parallel reactions to the methanol formation reaction, such as the formation of formaldehyde and ferric acid. In addition, the oxidation of methanol over longer electrolysis times may contribute to limiting the concentration values obtained.

Figure 39 shows the variation in formaldehyde concentration as a function of the experiment time, where it can be seen that no formaldehyde was formed up to 1.9 V vs. ECS. At more positive potentials, the concentration of formaldehyde increased as the applied potential increased, reaching a maximum concentration at 2.3 V vs. ECS with 19 mg L^{-1} of formaldehyde.

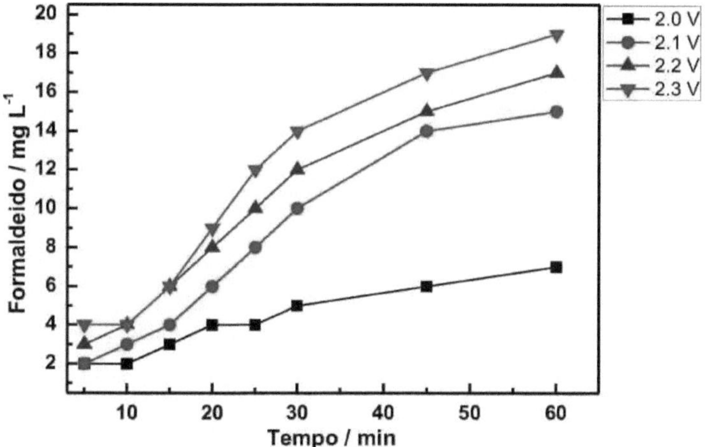

Figura 39 - Variation of formaldehyde concentration as a function of experiment time using EDG with 20% vanadium

Looking at Figure 39, it can also be seen that the variation in the concentration of formaldehyde also shows a stabilization profile in concentration after 30 minutes of reaction, probably due to the parallel process of oxidation of the compound, as observed by BOBROVA and collaborators, who mention various possible reactions during the catalytic oxidation of methane (Bobrova *et al*, 2007).

- EDG with 10% vanadium oxide

In the methane oxidation experiments using the EDG of $(TiO_2)_{0.679}(RuO_2)_{0.292}(V_2O_5)_{0.029}$, it was decided to quantify all the products identified in the methane oxidation process.

Figure 40 shows the variation in the concentration of methanol as a function of experimental time when using the EDG of $(TiO_2)_{0.679}(RuO_2)_{0.292}(V_2O_5)_{0.029}$, where the increase in the final concentration of methanol can be seen with the increase in the applied potential, reaching a maximum of 240 mg L^{-1} at 2.3 V vs. ECS.

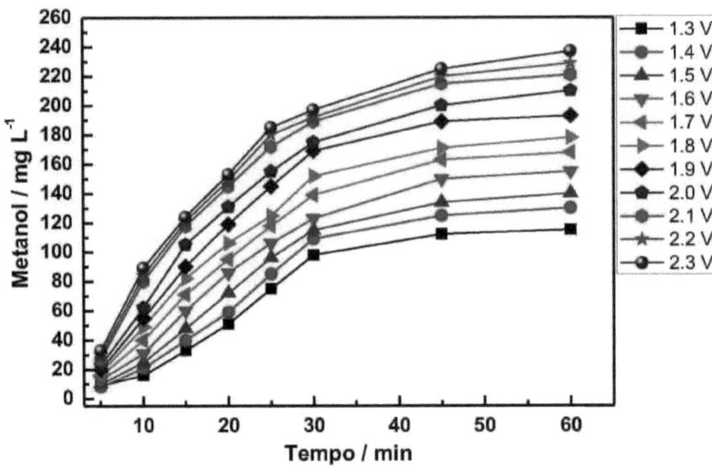

Figura 40 - Variation of methanol concentration as a function of experiment time using EDG with 10% vanadium. Electrolyte: 20 mL Na2SO4 0.1 mol L^{-1}

Figure 40 also shows a trend towards stabilization of the variation in methanol concentration as a function of experiment time from 30 minutes onwards. The same trend was observed in the experiments with EDG with 20% vanadium. This stabilization may be associated with parallel reactions occurring in the system (Bobrova *et al*, 2007).

Figure 41 shows the variation in the concentration of formaldehyde as a function of the experiment time. It can be seen that the detection of formaldehyde only occurs at potentials higher than 1.7 V vs. ECS, reaching a maximum concentration of 24 mg L^{-1} at 2.3 V vs. ECS after 1 hour of experimenting.

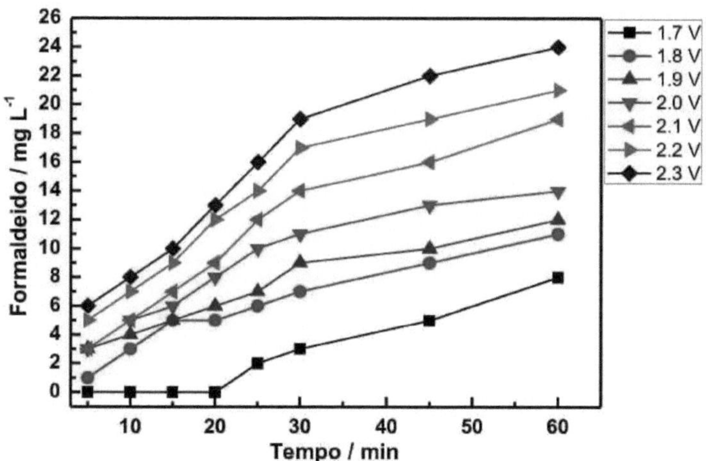

Figura 41 - Variation of formaldehyde concentration as a function of experiment time using EDG with 10% vanadium. Electrolyte: 20 mL Na2SO4 0.1 mol L^{-1}

Comparing Figures 39 and 41, it can be seen that as the amount of catalyst used in the EDG decreased, the maximum concentrations of formaldehyde increased, from 19 mg L^{-1} with 20% vanadium to 24 mg L^{-1} with 10% vanadium. Another point observed was the potential range with formaldehyde detection, with 20% vanadium formaldehyde was detected from 2.0 V vs. ECS and with 10% catalyst formaldehyde was detected from 1.7 V vs. ECS.

Another difference between the gas diffusion electrodes with 20% and 10% catalyst was the detection of ferric acid in the experiments with EDG catalyzed with 10% vanadium, which had not appeared using the electrode with 20% catalyst. The results of the quantification of ferric acid are shown in Figure 42.

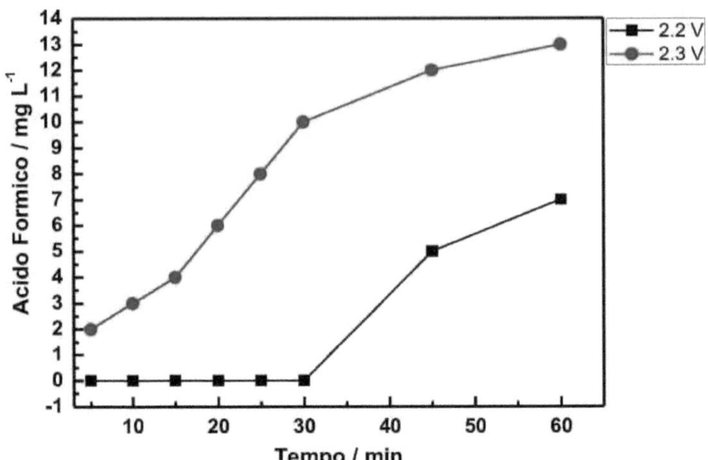

Figura 42 - Variation of the concentration of ferric acid as a function of time in an experiment using EDG with 10% vanadium. Electrolyte: 20 mL Na2SO4 0.1 mol L^{-1}

Figure 42 shows the variation in the concentration of ferric acid as a function of the experiment time. It can be seen that the detection of sorbic acid only occurred at the potentials of 2.2 V and 2.3 V vs. ECS, and at the potential of 2.2 V sorbic acid was only detected after 30 minutes of experimenting and the maximum concentration was reached, with EDG catalyzed with 10% vanadium, in the experiment at 2.3 V vs. ECS with 13 mg L .$^{-1}$

Comparing the results obtained with the electrodes catalyzed with 20% and 10% vanadium, it can be seen that with the reduction of the catalyst there was a decrease in the formation of methanol, but there was an increase in the formation of formaldehyde and the detection of ferric acid in the electrode with 10% vanadium. After analyzing the results presented, it was planned to use gas diffusion electrodes with 5% vanadium in order to study the influence of reducing the catalyst on the formation of methane oxidation products.

- EDG with 5% vanadium oxide

In the methane oxidation experiments using the $(TiO_2)_{0.689}(RuO_2)_{0.296}(V_2O_5)_{0.015}$ EDG, methanol, formaldehyde and thermal acid were detected, and the results are shown below.

The results with EDG catalyzed with 5% vanadium indicated that methanol generation was even lower compared to the experiments catalyzed with 20% and 10% vanadium oxide, at the highest potential applied (2.3 V vs. ECS) 200 mg L^{-1} was reached and at the lowest potential applied 90 mg L .$^{-1}$

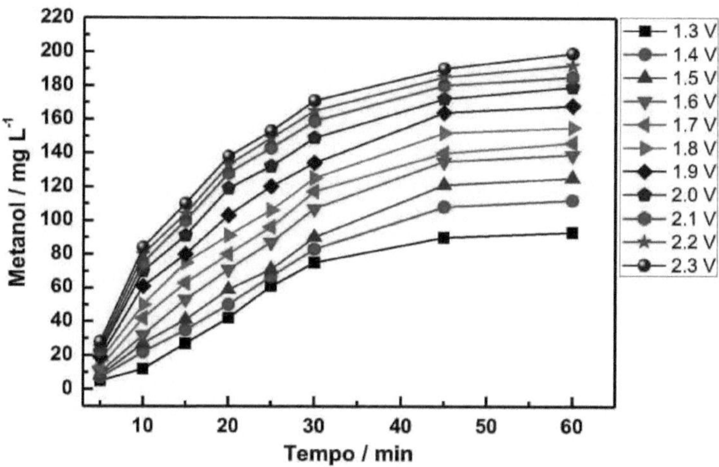

Figura 43 - Variation of methanol concentration as a function of experiment time using EDG with 5% vanadium. Electrolyte: 20 mL Na2SO4 0.1 mol L-1

Figure 43 shows the same trend of stabilization in the variation of the methanol concentration as a function of the experiment time from 30 minutes onwards, observed in the EDGs with 20% and 10% vanadium. As in the previous experiments, this stabilization may be associated with parallel reactions occurring in the system. Comparing the results of the electrodes with 20% and 10% vanadium, it can be seen that the reduction in catalyst also influenced the formation of formaldehyde, as can be seen in Figure 44.

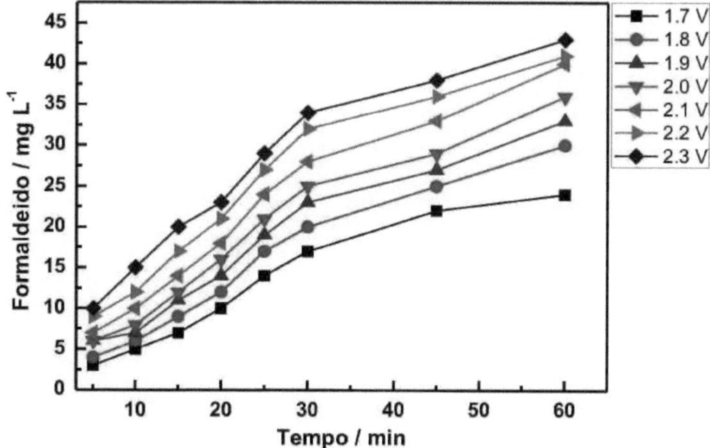

Figura 44 - Variation of formaldehyde concentration as a function of experiment time using EDG with 5% vanadium. Electrolyte: 20 mL Na2SO4 0.1 mol L^{-1}

Looking at Figure 44, it can be seen that with the increase in potential applied, there was an increase in the final concentrations in each experiment, reaching a maximum of 43 mg L^{-1} of formaldehyde in 1 hour of experimentation. As with the results for EDG with 20% and 10% vanadium, the formaldehyde formation profile did not show a clear stabilization trend, as observed in all the methanol results.

Analyzing the formaldehyde formation results shown in Figure 44, it can be seen that as the percentage of catalyst decreased, there was an increase in the final formaldehyde concentrations, but in the same potential range, 1.7 V to 2.3 V vs. ECS, when compared to the EDG catalyzed with 10% vanadium oxide. As the catalyst decreased, the detection of thermal acid was also observed, as can be seen in Figure 45.

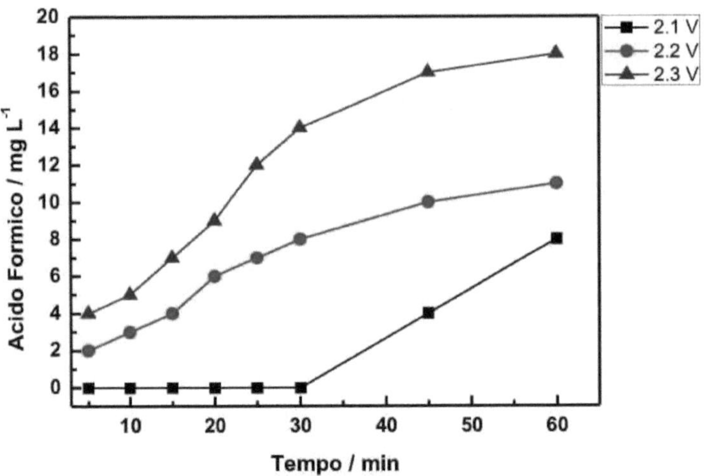

Figura 45 - Variation of the concentration of ferric acid as a function of time in the experiment using EDG with 5% vanadium. Electrolyte: 20 mL Na2SO4 0.1 mol L^{-1}

Figure 45 shows the variation in the concentration of the ferric acid as a function of the experiment time. It can be seen that the concentration of the acid increases with the increase in the applied potential, reaching 18 mg L^{-1} of ferric acid at the applied potential of 2.3 V vs. ECS in 1 hour of the experiment.

In Figures 42 and 45, it can be seen that the beginning of the detection of ferric acid occurs after 30 minutes, both in the GDE with 10% catalyst and in the GDE with 5% catalyst. It can also be seen that ferric acid was only detected at potentials 2.1V, 2.2V and 2.3V vs. ECS, unlike the GDE with 10%

vanadium with detection at 2.2V and 2.3V vs. ECS and the GDE with 20% vanadium, where ferric acid was not detected at any potential.

Analysis of the results of methane oxidation using EDG, catalyzed with 20%, 10% and 5% vanadium oxide, showed the formation of methanol, formaldehyde and ferric acid. Figures 46 to 48 show the variation in methane oxidation products at 2.2 V potential vs. ECS.

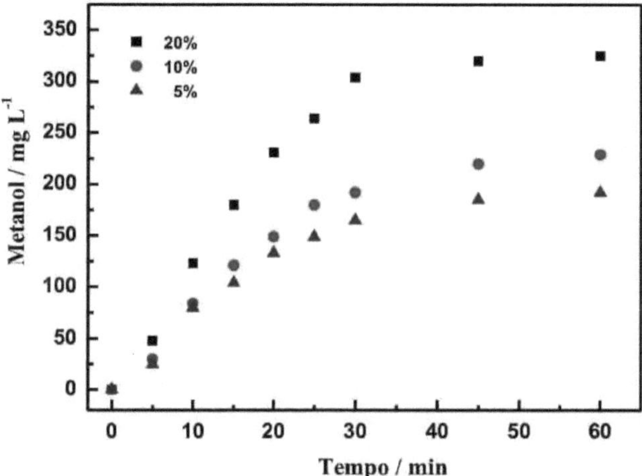

Figura 46 - Variation of the concentration of ferric acid as a function of time in an experiment using EDG with 5% vanadium at a potential of 2.2 V vs. ECS. Electrolyte: 20 mL Na2SO4 0.1 mol L^{-1}

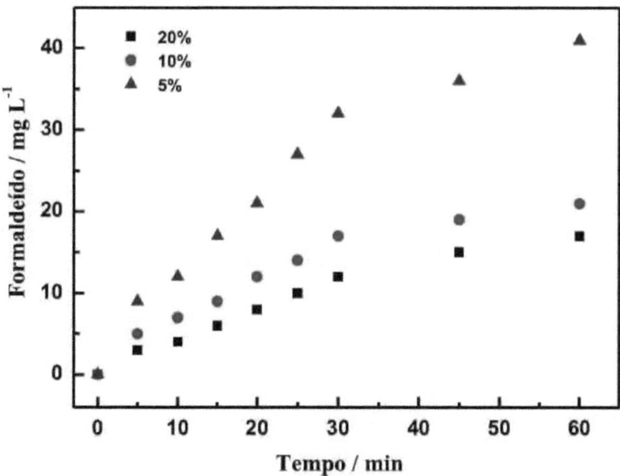

Figura 47 - Variation of the concentration of ferric acid as a function of time in an experiment using

EDG with 5% vanadium at a potential of 2.2 V vs. ECS. Electrolyte: 20 mL Na2SO4 0.1 mol L^{-1}

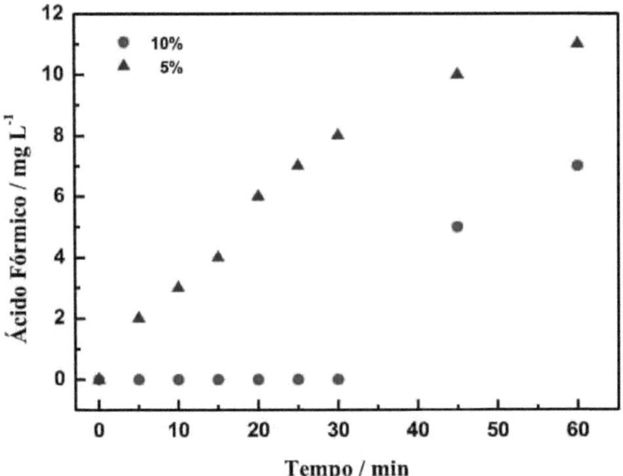

Figura 48 - Variation of the concentration of ferric acid as a function of time in an experiment using EDG with 5% vanadium at a potential of 2.2 V vs. ECS. Electrolyte: 20 mL Na2SO4 0.1 mol L^{-1}

With regard to the formation of methanol, the best electrode was catalyzed with 20% vanadium, reaching, according to Figure 46, 325 mg L^{-1} at a potential of 2.2 V vs. ECS. With regard to the formation of formaldehyde and ferric acid, the EDG with the best results was catalyzed with 5% vanadium, achieving 43 mg L-1 of formaldehyde (Figure 47) and 18 mg L^{-1} of ferric acid (Figure 48), both at 2.2 V vs. ECS.

Looking at Figure 46, it can be seen that an increase in the amount of catalyst led to an increase in the formation of methanol, but the same increase in the amount of catalyst led to a decrease in the concentrations of formaldehyde and ferric acid, as shown in Figures 47 and 48 respectively. These variations in the concentrations of methanol, formaldehyde and ferric acid may be associated with the variation in the amount of vanadium oxide, promoting the formation of methanol to the detriment of formaldehyde and ferric acid.

The results presented are promising when compared with the techniques and processes described in the literature, since the EDG catalyzed with 20% vanadium oxide obtained superior results, 10.4 x 10^{-3} mol L^{-1} of methanol, when compared with the results described in the literature.

using different techniques. KHOKHAR and colleagues achieved 2.7 x 10^{-3} mol L^{-1} of methanol and 2.4 x 10^{-3} mol L^{-1} of formaldehyde in 180 minutes of experiment using a volume of 100 mL with heterogeneous catalysis of a ruthenium complex. With regard to the formation of formaldehyde, the authors achieved better results, as the EDG catalyzed with 1% vanadium oxide reached 1.4 x 10^{-3} mol L^{-1} of formaldehyde (Khokhar *et al*, 2009). TABATA and OKURA used the bioprocess using *M. trichosporium* OB3b and achieved approximately 1.6 x10^{-1} mol L^{-1} of methanol in 120 minutes of experimentation (Tabata and Okura, 2008). RAZUMOVSKY *et al.* also used the bioprocess, but with the bacterium *M. sporium* B-2121 and achieved 8.5 x 10^{-3} mol L^{-1} of methanol in 180 hours of experimentation (Razumovsky *et al*, 2008).

With regard to the formation of ferric acid, PARK et al achieved 7.4 x 10^{-4} mol L^{-1} of the acid in an autoclave reactor with a volume of 300 mL catalyzed with $PdCl_2$, this result is superior to the EDG catalyzed with 1% vanadium oxide with 3.9 x 10^{-4} mol L^{-1} (Park *et al*, 2000).

When comparing the results of the catalyzed EDG with the results described in the literature, it was observed that the gas diffusion electrodes are efficient in oxidizing methane gas, but it is necessary to quantify the chemical efficiency in converting the methane molecules into methanol and the efficiency of applying the electrical charge to the methanol formation reaction.

Studying the chemical efficiency of methane oxidation and methanol formation is important in order to establish the minimum amount of gas that should be supplied to the EDG during the experiment, so that the consumption of this reagent during the synthesis process can be reduced as much as possible. The efficiency results of the EDGs with different amounts of catalyst are shown in Figure 49.

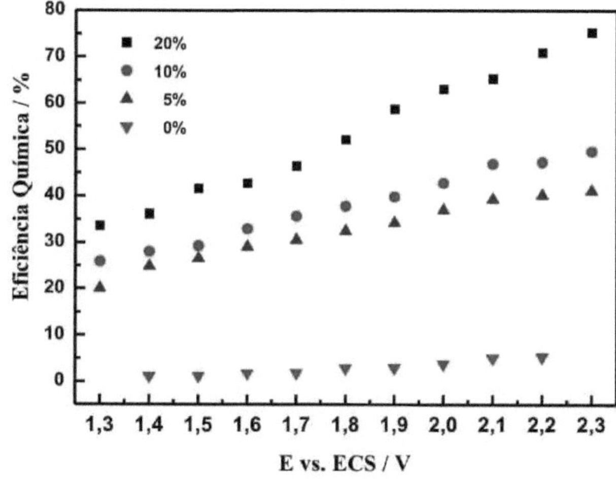

Figura 49 - Variation in the chemical efficiency of methanol formation as a function of the applied potential using EDG with and without vanadium oxide

Figure 49 shows the results for EDG with 20%, 10% and 5% vanadium. It can be seen that the chemical efficiency increases as the applied potential increases, with the lowest efficiency being 35% at a potential of 1.3V vs. ECS and the highest efficiency being 75% at a potential of 2.3V vs. ECS, for EDG catalyzed with 20% vanadium. In the experiments with EDG catalyzed with 10% vanadium, it was observed that as the catalyst concentration decreased, the chemical efficiency in the conversion of methane to methanol decreased, reaching a maximum of 50% at a potential of 2.3 V vs. ECS. In relation to the experiments with EDG catalyzed with 5% vanadium, a further decrease in efficiency values was observed, reaching a maximum of 42% at 2.3 V vs. ECS. The electrodes catalyzed with vanadium oxide obtained better results compared to the EDG without catalyst, where the best potential, 2.2 V vs. ECS, achieved 5.5% chemical efficiency for the methanol formation reaction.

Figures 49 show an increase in chemical efficiency with an increase in the concentration of catalyst in the EDG, but this increase may be associated with an improvement in the methane oxidation process and is not directly related to the volume of gas supplied to the EDG during the experiments, because the volumes of methane gas used during the experiments varied around 7.5 ±0.4 mL in 1 hour of experimentation, while the efficiency values increase with the increase in the applied potential and with the increase in the catalyst.

The chemical efficiency values presented can be considered promising when compared with the techniques and processes described in the literature. FOULDS and GRAY studied the oxidation of methane in a homogeneous gas phase. The authors achieved a maximum efficiency of 2% in the conversion of methane to methanol at a reaction temperature of 420°C in a tubular flow reactor (Foulds and Gray, 1995). RAJA and RATNASAMY also studied the conversion of methane to methanol in an autoclave reactor, achieving approximately 5% efficiency in the formation of methanol and 2% in the formation of formaldehyde in the same experiment (Raja and Ratnasamy, 1997). AOKI and colleagues used a flow reactor to study the direct conversion of methane to methanol, the best results showed an efficiency of 13% in the formation of methanol (Aoki *et al*, 1998). ZHANG et al. studied the conversion of methane into methanol and formaldehyde and showed that the quartz reactor formed methanol and formaldehyde with an efficiency of approximately 4.7% for both products (Zhang *et al*, 2002).

In comparison with the results presented, the chemical efficiency presented by the electrodes without catalyst is close to the values described in the literature, but when the vanadium oxide catalyst was used, the values presented were much higher than those described in the literature, showing the improvement in efficiency when associating the gas diffusion electrodes with vanadium oxide as a catalyst.

With regard to the chemical efficiency results, when EDG catalyzed was used, the same trend was observed as with the variation in methanol concentration (Figures 38, 40 and 43), where increasing the amount of catalyst increased methanol formation at all the potentials studied, thus increasing the catalyst led to an increase in chemical efficiency. The increase in methanol formation with an increase in the amount of catalyst is associated with an increase in efficiency in the process, since the greater the number of methane molecules oxidized to methanol, the higher the final alcohol concentration and the higher the chemical efficiency values.

When studying the oxidation of methane to methanol using gas diffusion electrodes, it is important to determine the amount of electrical charge applied exclusively to the methanol formation reaction. To this end, the electrical efficiency for the alcohol formation reaction was determined and the results are shown in Figure 50.

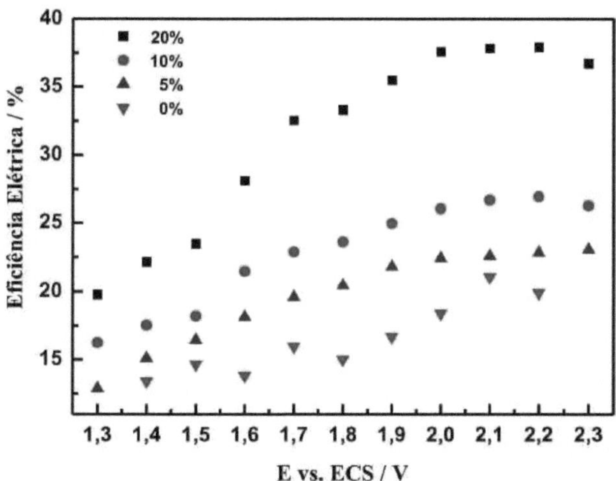

Figura 50 - Variation of electrical efficiency in methanol formation as a function of applied potential using EDG with and without vanadium oxide

Figure 50 shows the electrical efficiency values as a function of the applied potential for the three percentages of catalysts studied. The electrical efficiency values are calculated by comparing the total electrical charge of the experiment with the theoretical electrical charge required to form methanol, considering 2 electrons exchanged in the oxidation reaction.

It can be seen that the experiments with the EDG with 20% vanadium have the best electrical efficiency, at the potential of 2.0V vs. ECS with approximately 38% compared to the EDG without catalyst, but at the potential of 2.3V vs. ECS there is a decrease in efficiency, which may be associated with the better potential (2.1 V vs. ECS) for oxidizing methane to methanol.

Analyzing the results of the other electrodes (10% and 5% vanadium) shows a decrease in electrical efficiency as the amount of catalyst in the electrode decreases, reaching 26% and 22% at the best potentials using the electrodes with 10% and 5% vanadium, respectively. This decrease may be associated with the lower formation of methanol at the same potentials.

When methanol is formed by oxidizing methane, it is necessary to determine the amount of energy consumed by the system in the alcohol formation reaction. To do this, the energy consumption per kilo of methanol formed was determined and the results are shown in Figure 51.

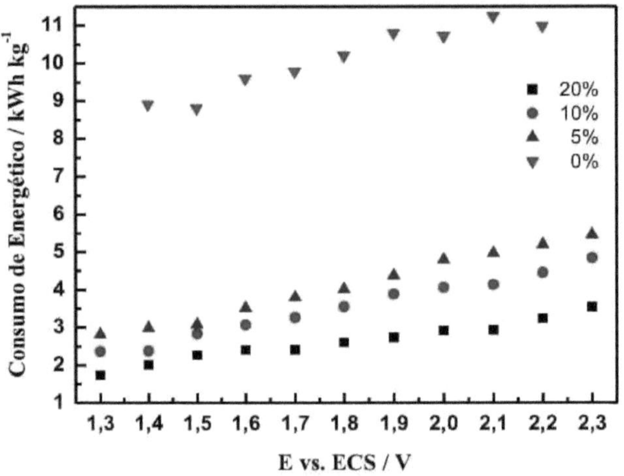

Figura 51 Variation in energy consumption per kilo of methanol formed as a function of applied potential using EDG with and without vanadium oxide

Figure 51 shows the values for energy consumption in the formation of methanol. It can be seen that an increase in the amount of catalyst leads to lower energy consumption in the formation of methanol. Energy consumption for the EDG with 20% vanadium is the lowest, reaching 3.5 kWh kg^{-1} of methanol, while the other electrodes show higher values at the same potentials applied, 4.9 kWh kg^{-1} and 5.5 kWh kg^{-1} for the electrodes with 10% and 5% vanadium, respectively. The higher values for energy consumption for the electrodes with 5% vanadium may be associated with the lower formation of methanol even though they provide approximate amounts of energy, i.e. all the electrodes were studied in the same potential range, but the EDG with 5% vanadium formed a smaller amount of methanol compared to the other electrodes. Analyzing the results of the experiments with catalyzed EDG, it can be seen that the addition of vanadium oxide led to a decrease in energy consumption for the formation of methanol, but in the experiments without catalyst, an increase in energy consumption was observed for the same reaction.

4.5 Ethylene oxidation using TiO_2RuO_2 EDGs

Analogous to the experiments on the oxidation of methane to form methanol, studies on the oxidation of ethylene gas aim to make use of the unique characteristic of metal oxide electrodes, which is the adsorption of atomic and radical species, To do this, it is possible to modulate the overpotential for the O_2 evolution reaction, preferably forming higher oxides, RuO_3 (Comninellis, 1997).

For the ethylene electroxidation experiments, a potential range of 1.3 V to 2.3 V vs. ECS was determined. The beginning of the potential range for the experiment was determined by observing the current profile of the O_2 release reaction (Figure 29) where the current increases at approximately 1.5 V vs. ECS and also by observing the results of the methane oxidation experiments, which showed an increase in methanol concentration and in the efficiency of the experiments as the applied potential increased. To determine the end of the potential range for the experiments, the same parameters were observed as for the methane experiments, i.e. the reduction in efficiency compared to the efficiency of the previous potential would define the limit potential for the experiments. For the experiments using ethylene, an approximate pressure of 0.04 Bar was determined. This pressure is necessary for small bubbles to form on the surface of the electrode; excessive accumulation of bubbles on the surface of the electrode could cause a reduction in the active surface of the EDG. To analyze the samples, a standard sample of ethylene glycol (the main product expected in the experiments) was inserted into the GC-Ms and the mass spectrum is shown in Figure 52.

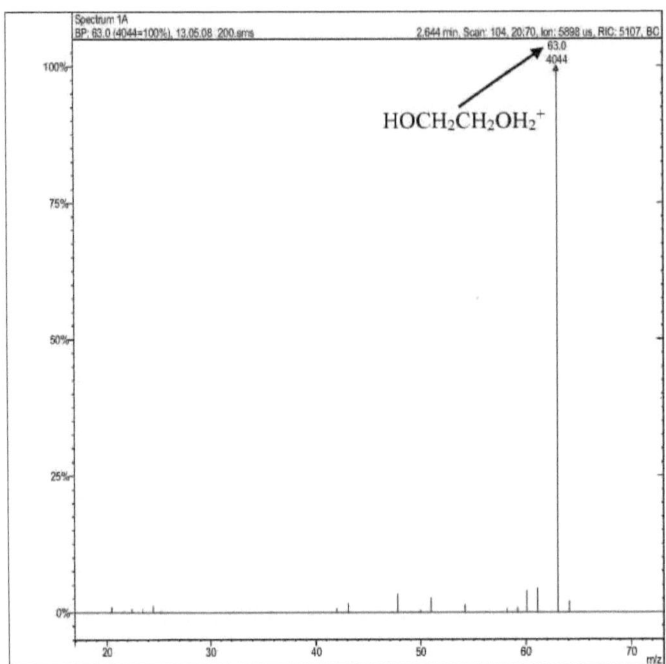

Figure 52 - Mass spectrum of the standard ethylene glycol sample

Figure 52 shows the result of inserting the standard ethylene glycol sample into the GC-Ms. The 63 *m/z* mass of the protonated ethylene glycol molecule can be seen; this phenomenon is characteristic of some molecules in the mass spectrometer *iontrap*.

Figure 53 shows the variation in the concentration of ethylene glycol as a function of the potential applied during the 1-hour experiment. It can be seen that the lowest concentration was approximately 12 mg L^{-1} at the potential of 1.3 V vs. ECS. As the potential applied increased, an increase in the concentration of methanol was observed, reaching a maximum value of approximately 161 mg L^{-1} at 2.3 V vs. ECS.

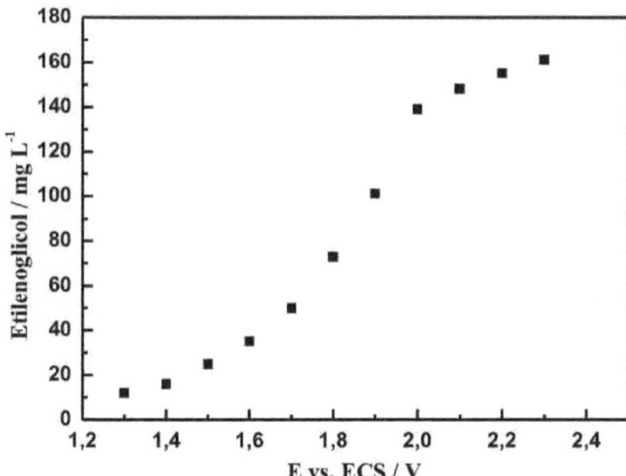

Figure 53 - Variation of ethylene glycol concentration as a function of applied potential. Potential range 1.3 V to 2.3 V vs. ECS. Electrolyte: 20 mL of 0.1 M Na2SO4

Figure 53 also shows a stabilization profile of the rate of formation of ethylene glycol as a function of the potential applied. It can be seen that up to the potential of 2.0 V vs. ECS the variation in formation is greater compared to the variations at higher potentials, reaching 161 mg L^{-1} of ethylene glycol at 2.0 V vs. ECS. This variation in the formation of ethylene glycol as a function of the potential applied may be associated with the best potential range for the reaction of ethylene oxidation to ethylene glycol.

The samples were inserted into a gas chromatograph coupled to a mass spectrometer. Figure 54 shows the mass spectrum of the sample from the experiment at 2.0 V vs. ECS, and shows the 63 m/z mass with an intensity of 7808 *counts*, referring to the protonated ethylene glycol molecule.

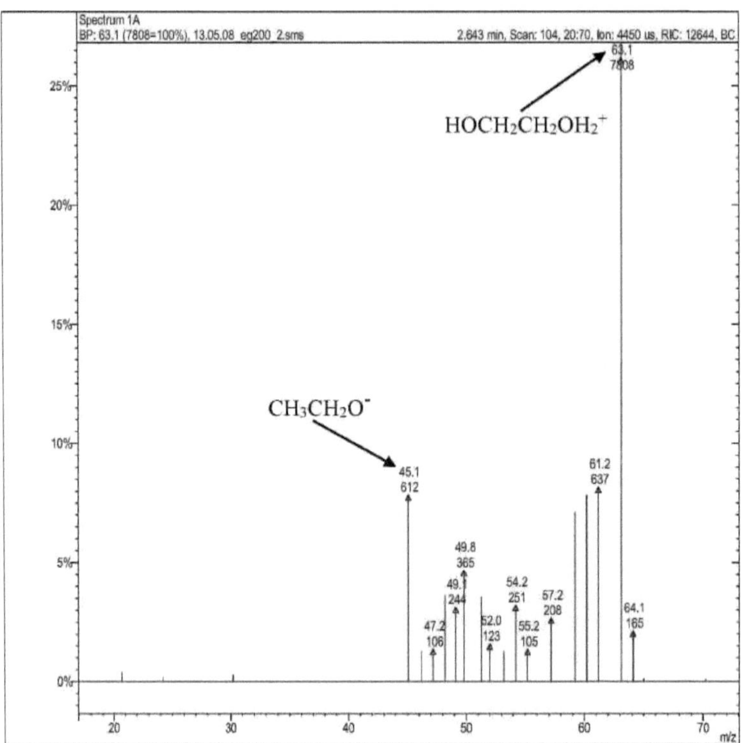

Figure 54 - Mass spectrum of the final sample of the experiment at 2.0 V vs. ECS

Looking at Figure 54, you can see another mass, with a value of 45 *m/z*, possibly associated with the deprotonated ethanol molecule. To confirm this molecule, a sample of standard ethanol was inserted and the corresponding mass spectrum is shown in Figure 55.

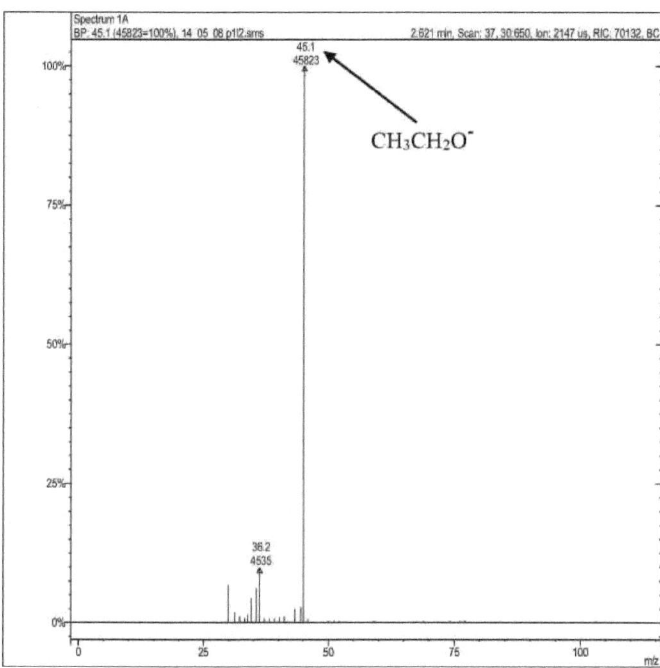

Figure 55 - Mass spectrum of the standard ethanol sample

As can be seen in Figure 55, the standard ethanol sample generated an ion of mass 45 m/z referring to the deprotonated ethanol molecule. Confirmation of the structure using the Ms/Ms technique, successive fragmentations of the molecular structure, could not be carried out because the fragments formed would have very small masses, which would make it impossible to read the results correctly.

Figure 55 shows the variation in the concentration of ethylene glycol and ethanol as a function of the potential applied during the 1-hour experiment. It can be seen that the lowest concentration of ethanol was approximately 12 mg L^{-1} at a potential of 1.3 V vs. ECS. As the potential applied increased, there was an increase in the concentration of ethanol, reaching a maximum value of approximately 41 mg L^{-1} at 1.7 V vs. ECS.

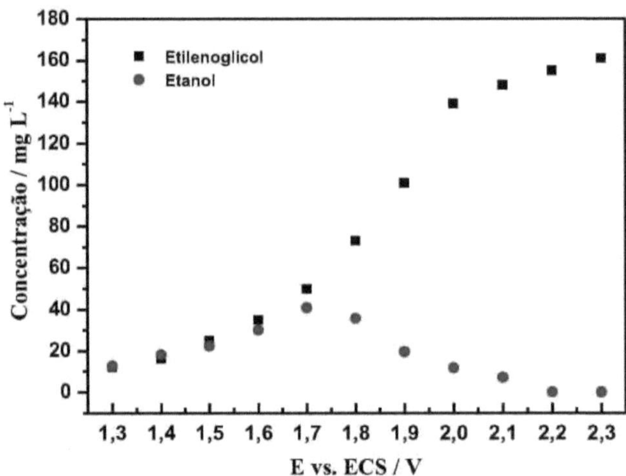

Figura 56 - Variation of ethylene glycol and ethanol concentration as a function of applied potential. Potential range 1.3 V to 2.3 V vs. ECS. Electrolyte: 20 mL of 0.1 M Na2SO4

It can also be seen in Figure 56 that at potentials above 1.7 V vs. ECS, the concentration of ethanol decreases with the increase in potential applied and at potentials of 2.2 V and 2.3 V vs. ECS, the 45 m/z ion is not observed in the samples analyzed, indicating that the concentration is close to zero or below the detection limit of the equipment. This decrease in ethanol concentration at potentials above 1.7 V vs. ECS may be associated with the instability of ethanol in the reaction medium, its anodic degradation and/or the continuity of the oxidation reaction on the EDG surface.

With the quantification of ethylene glycol and ethanol in the electrolysis with ethylene in the oxide EDG, it was necessary to determine the chemical efficiency in the modification of ethylene to ethylene glycol and ethanol. For this calculation, it was necessary to determine the amount of ethylene supplied to the reaction through the electrode structure, with this information it is possible to know the total amount of gas supplied to the system as a function of a given pressure.

Equation 14 was used to calculate the ethylene flow rate, where the gas pressure during the experiment was used to calculate the equivalent volume of ethylene. With the volume defined, the general law of gases under normal environmental conditions of temperature and pressure (CNATP: T=20 °C and P=1 atm) was used to calculate the number of mois equivalent to the calculated volume of gas.

With the number of moles of ethylene defined, it was compared with the number of moles of ethylene glycol formed, this comparison determined the rate of ethylene glycol formed from the ethylene supplied, the results of the chemical efficiency are shown in Figure 57.

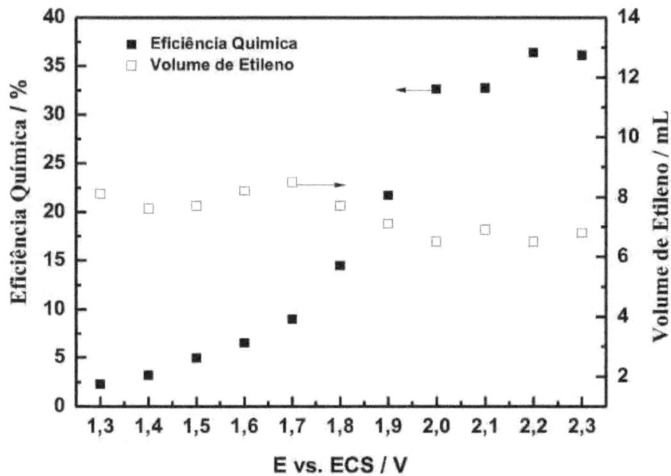

Figura 57 - Variation in the chemical efficiency of ethylene glycol compared to the volume of ethylene supplied as a function of the applied potential

Figures 57 show the variations in the chemical efficiency values of ethylene glycol as a function of the potential applied. It can be seen that the efficiency value of ethylene glycol reached approximately 36 % in the conversion of ethylene into ethylene glycol at 2.2 V vs. ECS.

Analyzing Figures 57, it can be seen that from 2.2 V vs. ECS the chemical efficiency of ethylene glycol, there is a small decrease in the efficiency value at potential 2.3 V vs. ECS compared to potential 2.2 V vs. ECS, possibly related to small variations in the gas supply (ethylene pressure adjustment) for this electrolysis.

It can also be seen in Figures 57 that the increase in chemical efficiency is associated with an increase in the potential applied and not related to the variation in the volume of gas during the experiments, because the volume of ethylene supplied showed small variations between 7 and 8 mL per hour of experiment, while the efficiency values increased with the increase in the potential applied, in the case of ethylene glycol.

Another point to be observed in the ethylene gas oxidation experiments is the efficiency of the electrical charge transfer for the ethylene oxidation reaction forming ethylene glycol. This electrical charge was compared with the total electrical charge values of the experiment (data provided by the potentiostat management program) and the results are shown in Figure 58.

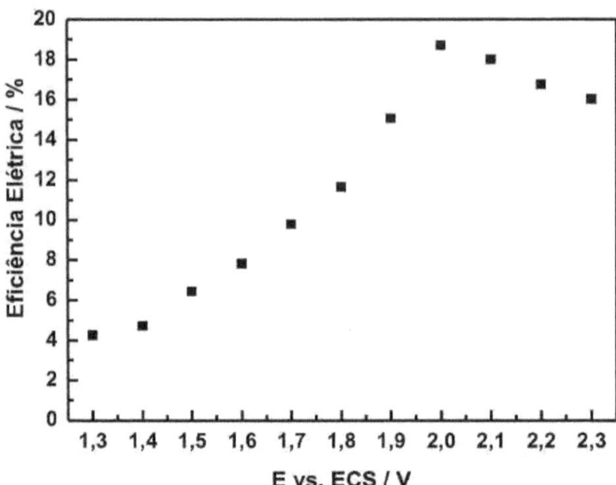

Figura 58 - Variation of the Electrical Efficiency (%) of ethylene glycol as a function of the applied potential

Figure 58 shows the variation in electrical efficiency as a function of the potential applied to ethylene glycol. It can be seen that the efficiency increases with the increase in applied potential up to 2.0 V vs. ECS with 18% for the ethylene glycol formation reaction. Looking at the ethylene oxidation reaction for ethylene glycol, there was a decrease in electrical efficiency from 2.0 V vs. ECS, which may be associated with the better potential for ethylene oxidation even with the increase in ethylene glycol concentration up to 2.3 V vs. ECS (Figure 56). ECS (Figure 56), indicating that even when applying potentials with more positive values, the reaction was less efficient in applying the electrical charge for the ethylene glycol formation reaction.

In relation to the application of the electric charge, another point to be observed is the consumption of electrical energy involved in the ethylene oxidation reaction, Figure 59 shows the energy consumption as a function of the applied potential.

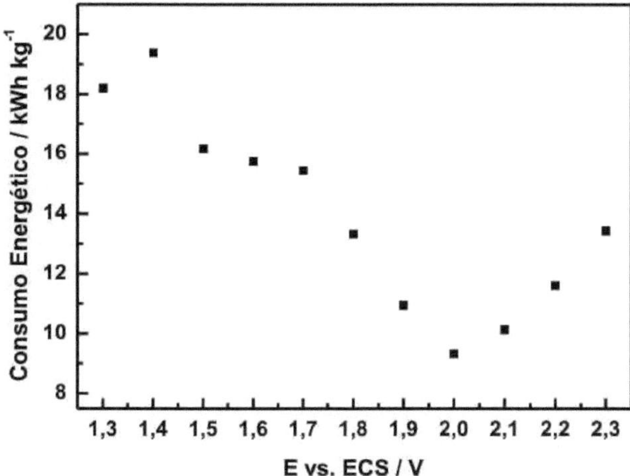

Figura 59 - Variation in Energy Consumption (kWh kg^{-1}) of ethylene glycol as a function of the applied potential

Figure 59 shows the variation in energy consumption as a function of the potential applied in the ethylene oxidation experiments. It can be seen that energy consumption decreases as the applied potential increases, reaching a minimum of approximately 9 kWh kg^{-1} of ethylene glycol formed and at more positive potentials an increase in energy consumption was observed, reaching a maximum of 14 kWh kg^{-1} of ethylene glycol at 2.3 V vs. ECS. Small variations can be seen in the energy consumed for the ethylene glycol formation reaction. This may be associated with the assembly of the electrochemical cell, because between each experiment, the electrochemical cell was disassembled for cleaning and replacement of the electrolyte, and when assembling the cell for the next experiment, it was not possible to position the electrodes identically to the previous experiment, thus interfering with the ohmic drop between each experiment.

Observing the results of the electrolysis with ethylene, we can see an increase in the concentration of ethylene glycol and ethanol as a function of the potential applied. However, in these electrolysis experiments, samples were only taken at the end of the experiment, making it impossible to analyze the variation in the concentration of ethylene glycol and ethanol during the electrolysis. Due to this impossibility, the sampling plan for the ethylene gas oxidation experiments was altered, with new samples being taken every 5 minutes for the first 30 minutes and every 15 minutes for the last 30 minutes. The aim of this change was to analyzc the variation in product concentration throughout the experiment.

Analyzing the results shown in Figures 56, 57 and 58, we can see an increase in the concentration of ethylene glycol and in the chemical and electrical efficiency as the applied potential increases, but we can see that the electrical efficiency is lower than the chemical efficiency in the generation of ethylene glycol. In order to better study the variations in chemical and electrical efficiency, constant current experiments were planned, using current densities of 0.15 mA cm^{-2} , 0.40 mA cm^{-2} , 0.70 mA cm^{-2} , 1.50 mA cm^{-2} , 5.00 mA cm^{-2} , 10.00 mA cm^{-2} , 20.00 mA cm^{-2} . The current density values were established from the limit current value, 0.25 mA, of the experiment with the highest electrical efficiency, 2.0 V vs. ECS, this limit current value was converted to current density, 0.40 mA cm^{-2} and to complete the sequence of experiments the other current density values were established. The results of the constant current experiments are shown in Figure 60.

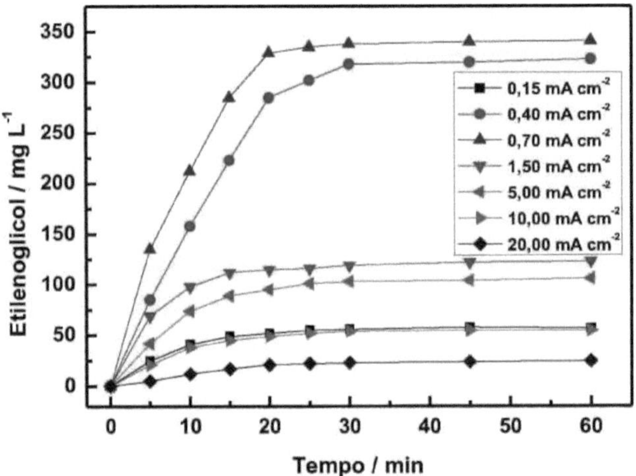

Figura 60 - Variation of ethylene glycol concentration as a function of applied current. Electrolyte: 20 mL Na2SO4 0.1 mol L^{-1}

As can be seen in Figure 60, all the electrolysis experiments had the same ethylene glycol concentration variation profile, with an increase in ethylene glycol concentration at the start of the experiment up to 30 minutes, followed by a stabilization in the concentration variation until the end of the experiments.

Figure 60 shows an increase in the concentration of ethylene glycol with the increase in the current density applied up to 0.7 mA cm^{-2} , from this potential the concentration decreases with the increase in the current density applied; this variation can be better observed in Figure 61.

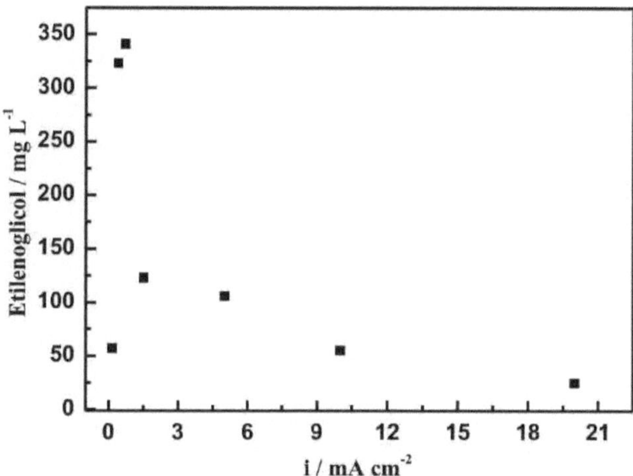

Figura 61 - Variation of the final concentration of ethylene glycol as a function of the current density applied. Electrolyte: 20 mL Na2SO4 0.1 mol L^{-1}

Figure 61 shows the variation in the final concentration of ethylene glycol as a function of the current density applied. There is an increase in the final concentration of ethylene glycol with the increase in the current applied up to a current of 0.7 mA cm^{-2} (341 mg L^{-1} of ethylene glycol), at higher currents there was a decrease in the concentration of ethylene glycol with the increase in the current density applied, reaching the lowest value at 20.00 mA cm^{-2} with a concentration of 25 mg L^{-1} of ethylene glycol.

This decrease in the concentration of ethylene glycol at currents above 0.7 mA cm^{-2} may be associated with the excess current density applied in relation to the volume of gas and the active area of the EDG, so the excess current density could be generating other reactions parallel to the formation of ethylene glycol, such as the o2 release reaction. The decrease in ethylene glycol formation with increasing current density can also be seen in the variation in chemical efficiency, Figure 62.

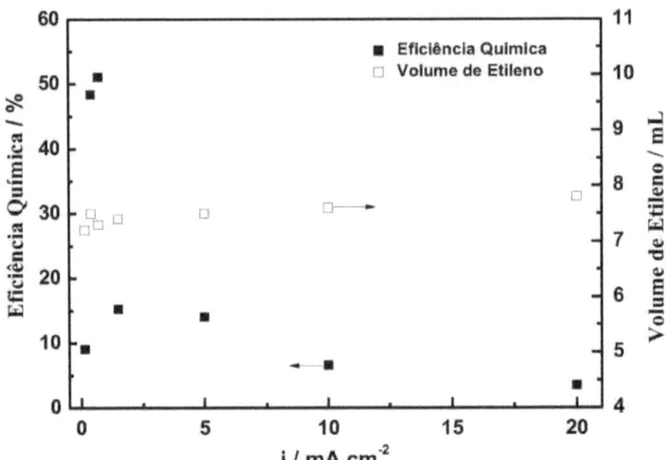

Figura 62 - Variation in chemical efficiency (%) and the volume of ethylene supplied (mL) both as a function of the current density applied.

Figure 62 shows the variation of the chemical efficiency values as a function of the applied potential. It can be seen that the values reached a maximum of approximately 51 % at a current density of 0.7 mA cm^{-2} . It can also be seen that the increase in chemical efficiency is associated with the increase in applied current density and not related to the variation in gas volume during the experiments, because the volume of ethylene supplied was 7.6 ±0.3 mL per hour of experiment, while the efficiency values increase with the increase in applied current density up to 0.7 mA cm .$^{-2}$

Another point to be observed in the ethylene gas oxidation experiments is the efficiency of the electrical charge transfer for the ethylene oxidation reaction to form ethylene glycol. The electrical charge required to form the appropriate final concentrations of ethylene glycol was calculated (Figure 60), This specific electric charge of ethylene glycol was compared with the total electric charge values of the experiment (data provided by the potentiostat management program) and the results are shown in Figure 63.

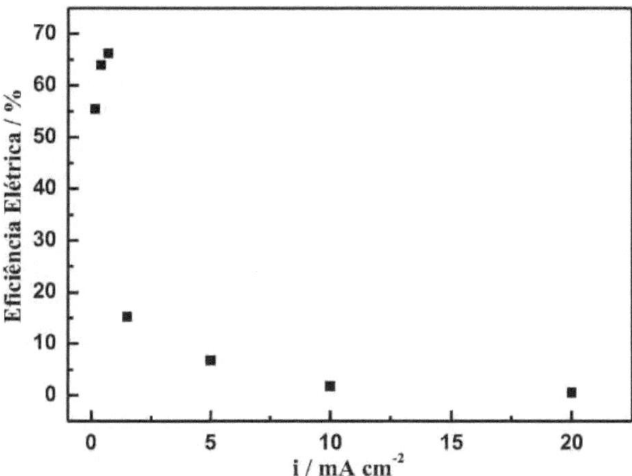

Figura 63 - Variation in Electrical Efficiency (%) as a function of applied current

Figure 63 shows the variation in electrical efficiency as a function of the current density applied. It can be seen that the efficiency increases as the current density applied increases up to 0.7 mA cm^{-2} , reaching an efficiency of approximately 66%. At higher current densities, there was a decrease in electrical efficiency, reaching a minimum value of 0.5% at 20.00 mA cm .$^{-2}$

This decrease may be associated with the better current density of 0.7 mA cm^{-2} for ethylene oxidation and at higher current densities the formation of ethylene glycol decreases, thus imposing a rapid decrease in efficiency at higher current densities.

The electrical efficiency results show that the charge transfer for the oxidation reaction of ethylene to ethylene glycol is more efficient compared to the chemical transformation of the ethylene molecule into ethylene glycol, This may be another indication of the excess of gas passing through the EDG, even with the charge transfer at approximately 66% (0.7 mA cm^{-2}) the chemical conversion was 51%, indicating that even applying 66% of the total charge of the experiment, the EDG converted 51% of the ethylene molecules.

With regard to the application of the electrical charge, another point to be observed is the consumption of electrical energy involved in the oxidation reaction of ethylene to ethylene glycol, Figure 64 shows the energy consumption as a function of the current density applied.

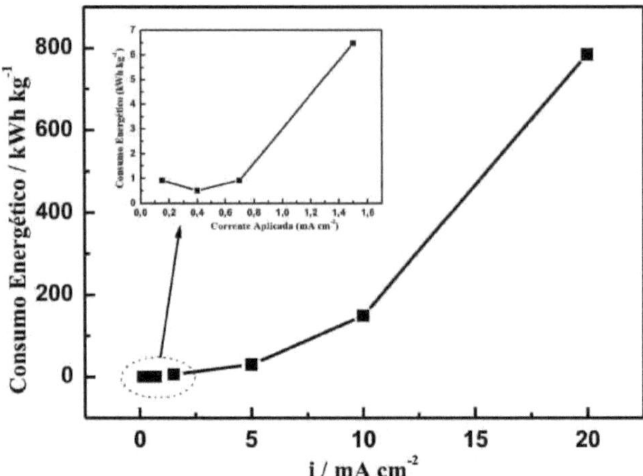

Figura 64 - Variation in energy consumption as a function of applied current density

Figure 64 shows the variation in energy consumption as a function of the current density applied in the experiments to oxidize ethylene to ethylene glycol. It can be seen that energy consumption increases as the current density applied increases, with a higher rate from 1.5 mA cm^{-2} as shown in Figure 64, reaching a maximum value of approximately 780 kWh kg^{-1} of ethylene glycol at 20 mA cm^{-2} with. Figure 64 shows the variation in energy consumption up to a current density of 1.5 mA cm^{-2} , showing that the lowest energy consumption was at 0.4 mA cm^{-2} with 0.5 kWh kg^{-1} followed by the currents 0.15 mA cm^{-2} and 0.7 mA cm^{-2} both with 0.9 kWh kg^{-1} of ethylene glycol formed in one hour of the experiment.

4.5.1 Ethylene oxidation using TiO2RuO2 EDG catalyzed with vanadium

Analysis of the results of ethylene gas oxidation showed the ability of the TiO2RuO2 EDG to form ethylene glycol and ethanol. The oxidation process showed a chemical efficiency of approximately 51% in the conversion of ethylene to ethylene glycol. With regard to the application of the electrical charge to the formation of ethylene glycol, the electrical efficiency reached approximately 66% and an energy consumption of 0.5 kWh kg^{-1} of ethylene glycol formed.

With the results presented, the aim was to add vanadium (in oxide form) to the TiO2RuO2 EDG in certain concentrations, in order to study the interference of the addition of the catalyst on the formation of the products and on the efficiency of ethylene glycol oxidation. The results of the experiments catalyzed with vanadium are presented below.

- EDG with 20% vanadium oxide

In the ethylene glycol oxidation experiments using the EDG of $(TiO_2)_{0.661}(RuO_2)_{0.283}(V_2O_5)_{0.056}$, ethylene glycol and ethanol were quantified following the calibration curves in Figures 50 and 54, respectively. The results of ethylene oxidation catalyzed with 20% vanadium oxide are shown in Figures 65 and 66.

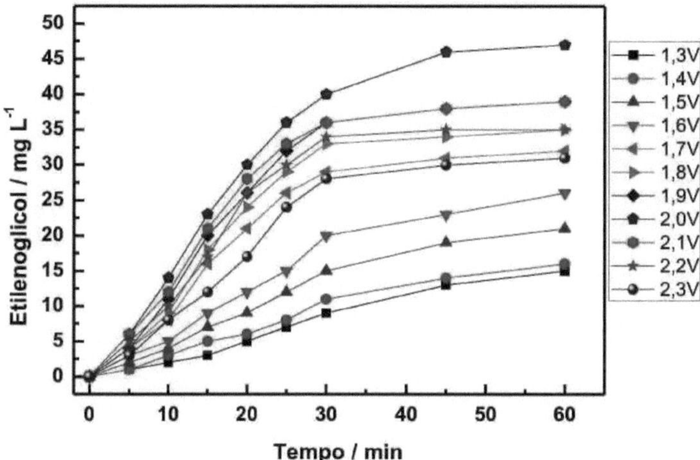

Figura 65 - Variation of ethylene glycol concentration as a function of experiment time using EDG with 20% vanadium. Electrolyte: 20 mL Na2SO4 0.1 mol L^{-1}

Figure 65 shows the variation in the concentration of ethylene glycol as a function of the experiment time. It can be seen that the increase in the potential applied led to an increase in the final concentration of ethylene glycol up to the potential of 2.0 V vs. ECS, reaching 47 mg L^{-1} and at more positive potentials there was a decrease in the final concentration of ethylene glycol, reaching 31 mg L^{-1} at 2.3 V vs. ECS. It can also be seen in Figure 65 that there is a rapid increase in ethylene glycol up to approximately 30 minutes into the experiment, after which there is a stabilization profile of the concentration as a function of the experiment time. Ethanol was also quantified alongside ethylene glycol and the results are shown in Figure 66.

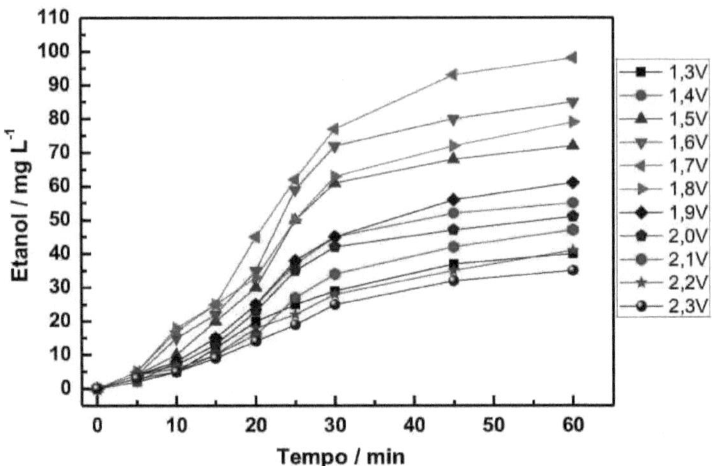

Figura 66 - Variation of ethanol concentration as a function of experiment time using EDG with 20% vanadium. Electrolyte: 20 mL Na2SO4 0.1 mol L^{-1}

Figure 66 shows the variation in ethanol concentration as a function of experiment time. It can be seen that alcohol concentrations change rapidly in the first thirty minutes of the experiment, showing a stabilization profile in the final thirty minutes of the experiment; this stabilization process was also observed in the quantification of ethylene glycol (Figure 65).

Figure 66 shows that increasing the applied potential led to an increase in the final concentrations of ethanol up to the potential of 1.7 V vs. ECS, reaching 98 mg L^{-1} . At more positive potentials, increasing the applied potential led to a decrease in the final concentrations, where the experiment at 2.3 V vs. ECS reached 35 mg L^{-1} at the end of the one-hour experiment. This variation in the final concentration of ethylene glycol and ethanol as a function of the applied potential can be better observed in Figure 67.

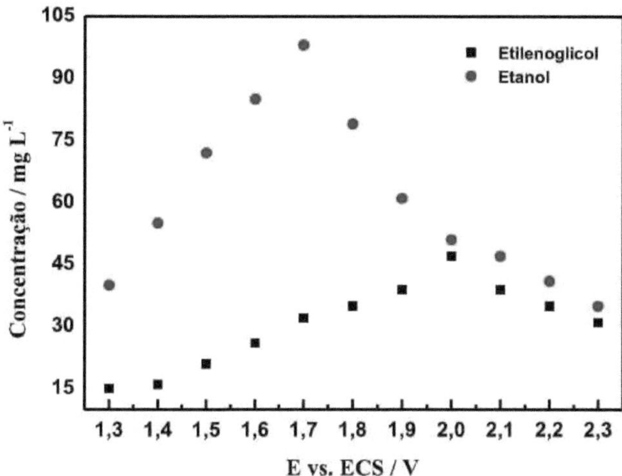

Figura 67 - Variation of the final concentration of ethylene glycol and ethanol as a function of the applied potential using EDG with 20% vanadium

Figure 67 compares the variation in the final concentration of ethylene glycol and ethanol as a function of the potential applied. It can be seen that increasing the potential applied led to an increase in the final concentrations of the products, but ethanol reached higher concentrations, 98 mg L^{-1} at 1.7 V vs. ECS, while ethylene glycol reached its maximum concentration at 2.0 V vs. ECS with 47 mg L^{-1} . At more positive potentials, there was a decrease in the final concentration of the two products, possibly associated with the better potential range for ethylene oxidation for each of the products.

- EDG with 10% vanadium oxide

Figures 68 and 69 show the results of the ethylene oxidation experiments using $(TiO_2)_{0.679}(RuO_2)_{0.292}(V_2O_5)_{0.029}$ EDG, with 10% vanadium oxide.

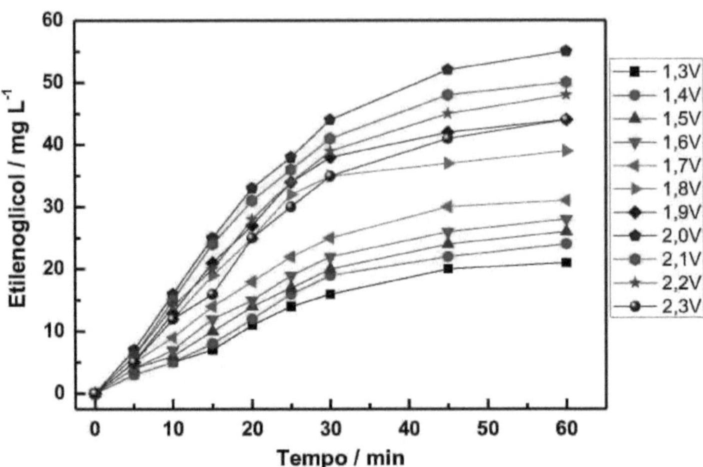

Figura 68 - Variation of ethylene glycol concentration as a function of experiment time using EDG with 10% vanadium. Electrolyte: 20 mL Na2SO4 0.1 mol L^{-1}

Figure 68 shows the variation in ethylene glycol concentration as a function of experiment time. It can be seen that, as in the case of EDG catalyzed with 20% vanadium oxide, there was an increase in the concentration of ethylene glycol as the applied potential increased up to 2.0 V vs. ECS with 55 mg L^{-1} , at more positive potentials there was a decrease in the concentration of ethylene glycol as the applied potential increased, reaching 44 mg L^{-1} at 2.3 V vs. ECS. Figure 68 also shows a tendency for the concentration to stabilize as a function of time from 30 minutes onwards. This process was also observed in the experiments with 20% catalyst and may be associated with the parallel reactions that occur concomitantly with the ethylene glycol formation reaction. In the experiments with EDG with 10% catalyst, ethanol was also quantified and the results are shown in Figure 69.

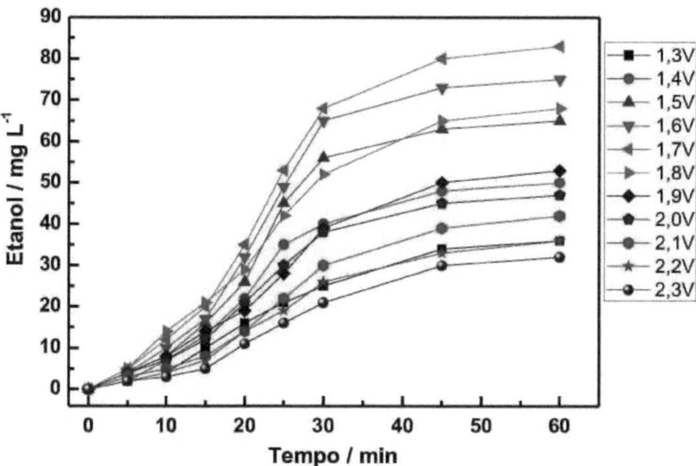

Figura 69 - Variation of ethanol concentration as a function of experiment time using EDG with 10% vanadium. Electrolyte: 20 mL Na2SO4 0.1 mol L^{-1}

Figure 69 shows the variation in ethanol concentration as a function of experiment time. It can be seen that increasing the applied potential led to an increase in ethanol concentration, reaching 83 mg L^{-1} at 1.7 V vs. ECS. At more positive potentials there was a decrease in the concentration of ethanol, reaching 32 mg L^{-1} of alcohol at 2.3 V vs. ECS. It should also be noted that this variation in concentration was also observed in the experiments catalyzed with 20% vanadium oxide. The variation in the concentration of ethylene glycol and ethanol as a function of the potential applied can be seen in Figure 70.

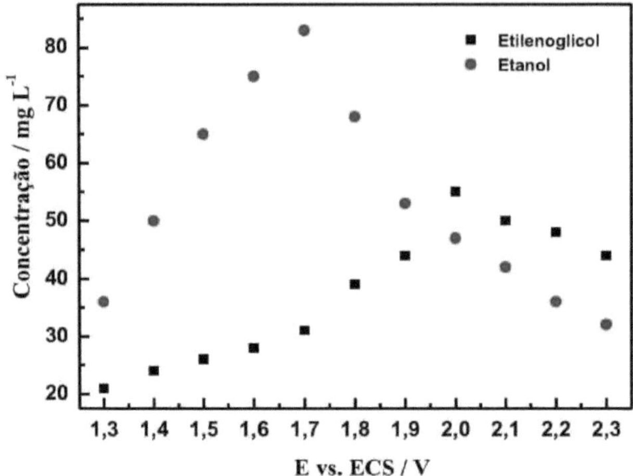

Figura 70 - Variation of the final concentration of ethylene glycol and ethanol as a function of the applied potential using EDG with 10% vanadium

Figure 70 compares the variation in the final concentration of ethylene glycol and ethanol as a function of the potential applied. It can be seen that increasing the potential applied led to an increase in the final concentrations of the products, but ethanol reached higher concentrations, 83 mg L^{-1} at 1.7 V vs. ECS, while ethylene glycol reached its maximum concentration at 2.0 V vs. ECS with 55 mg L^{-1} . At more positive potentials, there was a decrease in the final concentration of the two products, possibly associated with the better range of ethylene oxidation potentials for each of the products. In the experiments catalyzed with 10% vanadium oxide, it was observed that at the most positive potentials (above 2.0 V vs. ECS) the concentrations of ethylene glycol were higher than the concentrations of ethanol, as observed in the experiments with 20% catalyst; this change may be associated with the decrease in catalyst concentration.

- EDG with 5% vanadium oxide

For the ethylene gas oxidation experiments, EDG of $(TiO_2)_{0.689}(RuO_2)_{0.296}(V_2O_5)_{0.015}$ with 5% vanadium oxide was used and the results are presented below.

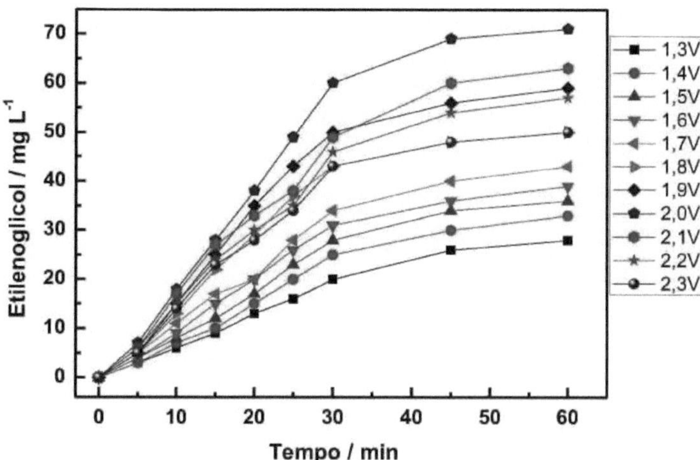

Figura 71 - Variation of ethylene glycol concentration as a function of experiment time using EDG with 5% vanadium. Electrolyte: 20 mL Na2SO4 0.1 mol L^{-1}

Figure 71 shows the variation in ethylene glycol concentration as a function of experiment time. It can be seen that the concentration increases with the increase in potential applied up to 2.0 V vs. ECS, reaching 71 mg L^{-1} of ethylene glycol. At more positive potentials, the concentration decreases, reaching 50 mg L^{-1} at 2.3 V vs. ECS. We can also observe a tendency for the concentration to stabilize as a function of time from approximately 30 minutes into the experiment; this process was also observed in the experiments catalyzed with 20% and 10%.

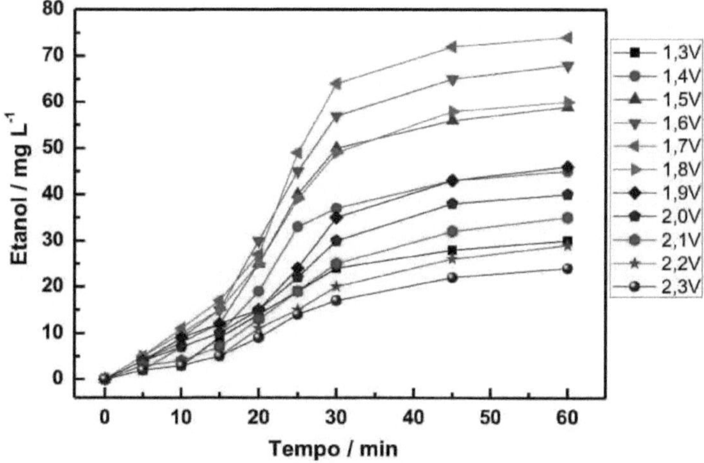

Figura 72 - Variation of ethanol concentration as a function of experiment time using EDG with 5% vanadium. Electrolyte: 20 mL Na2SO4 0.1 mol L^{-1}

Figure 72 shows the variation in ethanol concentration as a function of experiment time. It can be seen that increasing the applied potential led to an increase in ethanol concentration, reaching 74 mg L^{-1} at 1.7 V vs. ECS. At more positive potentials there was a decrease in concentration, reaching 24 mg L^{-1} of ethanol at 2.3 V vs. ECS. It should also be noted that this variation in concentration was also observed in the experiments catalyzed with 20% and 10% vanadium oxide, which can be considered a trend in the formation of ethanol from the oxidation of ethylene. There was also a tendency for the concentration of ethanol to stabilize after thirty minutes of experiments, and this trend was also observed in the experiments with different amounts of catalyst.

The variation in the concentration of ethylene glycol and ethanol using EDG catalyzed with 5% vanadium oxide can best be seen in Figure 73.

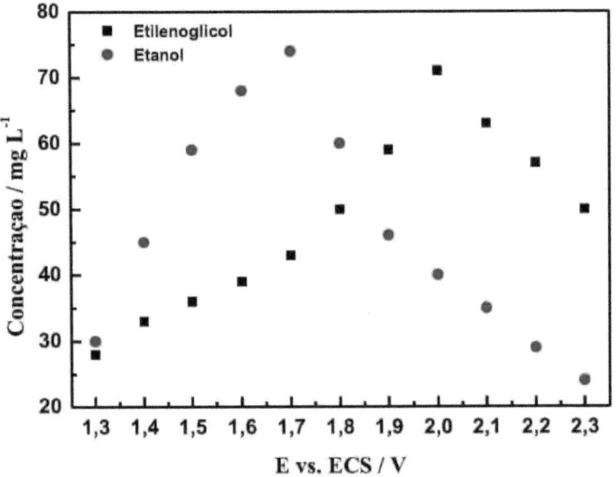

Figura 73 - Variation of the final concentration of ethylene glycol and ethanol as a function of the applied potential using EDG with 5% vanadium

Figure 73 compares the variation in the final concentration of ethylene glycol and ethanol as a function of the potential applied. It can be seen that increasing the potential applied led to an increase in the final concentrations of the products, but ethanol reached higher concentrations, 74 mg L^{-1} at 1.7 V vs. ECS, while ethylene glycol reached its maximum concentration at 2.0 V vs. ECS with 71 mg L^{-1} . At more positive potentials, there was a decrease in the final concentration of the two products, possibly associated with the better potential range for ethylene oxidation for each of the products.

In the experiments catalyzed with 5% vanadium oxide, it was observed that at the most positive potentials (above 2.0 V vs. ECS) ethylene glycol concentrations were higher than ethanol concentrations, as observed in the experiments with 20% and 10% catalyst, this change may be associated with the decrease in catalyst concentration. Figure 73 also shows that the maximum concentration of ethanol (1.7 V vs. ECS) is very close to the concentration of ethylene glycol (2.0 V vs. ECS). This proximity in concentration values was not observed in the experiments with different catalyst concentrations and may also be associated with a reduction in the amount of catalyst.

- EDG with 1% vanadium oxide

For the ethylene gas oxidation experiments, the $(TiO_2)_{0.698}(RuO_2)_{0.299}(V_2O_5)_{0.003}$ EDG was used with the smallest amount of catalyst, 5% vanadium oxide, and the results are presented below.

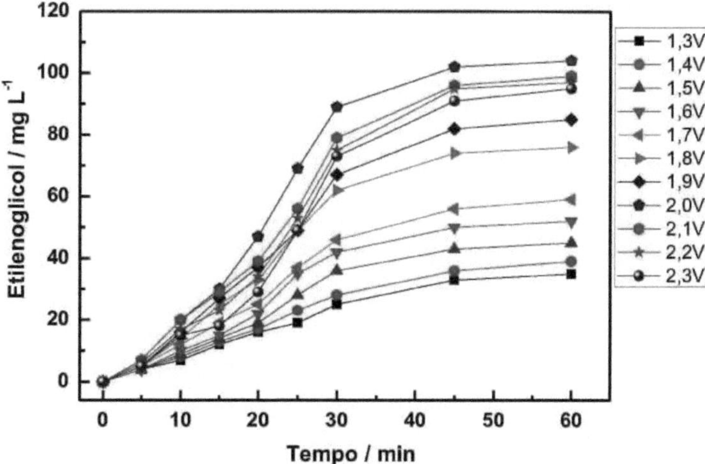

Figura 74 - Variation of ethylene glycol concentration as a function of experiment time using EDG with 1% vanadium. Electrolyte: 20 mL Na_2SO_4 0.1 mol L^{-1}

Figure 74 shows the variation in ethylene glycol concentration as a function of experimental time. There is a tendency for the concentration to stabilize as a function of time from 30 minutes onwards. This tendency may be associated with the parallel reactions that occur during ethylene oxidation.

Figure 74 also shows that the concentration of ethylene glycol increased with the increase in potential applied up to 2.0 V vs. ECS reaching 104 mg L^{-1} of ethylene glycol, at more positive potentials a decrease in the final concentration of the experiments was observed, this decrease was observed in all the experiments catalyzed with vanadium oxide, but it was not observed in the experiments with TiO_2RuO_2 EDG where the concentration of ethylene glycol increased with the increase in potential applied up to 2.3 V vs. ECS. In the experiments catalyzed with 1% vanadium oxide, ethanol was also quantified and the results are shown in Figure 75.

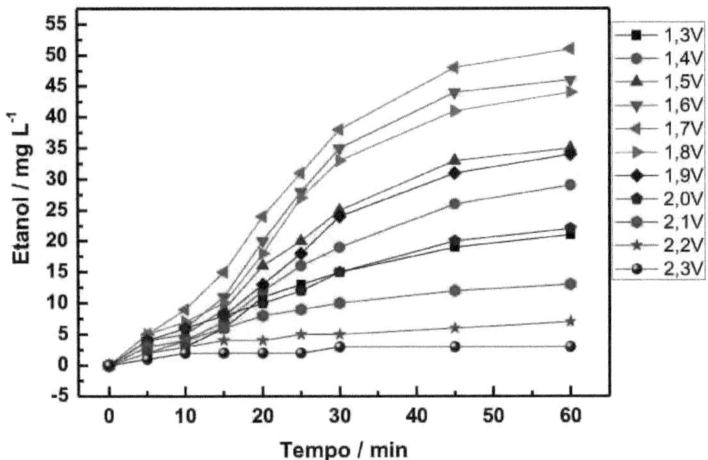

Figura 75 - Variation of ethanol concentration as a function of experiment time using EDG with 1% vanadium. Electrolyte: 20 mL Na2SO4 0.1 mol L^{-1}

Figure 75 shows the variation in ethanol concentration as a function of experiment time. It can be seen that the concentration of ethanol increases with the increase in potential applied up to 1.7 V vs. ECS, reaching 51 mg L^{-1} of ethanol; at more positive potentials it was observed that the concentrations decreased, reaching 3 mg L^{-1} of ethanol at 2.3 V vs. ECS. Figure 75 also shows a tendency for the concentration of ethanol to stabilize as a function of time from 30 minutes onwards. This tendency was observed in all the experiments catalyzed with vanadium oxide.

The results of ethylene oxidation and ethanol formation reached a maximum of 98 mg L^{-1} of alcohol at 1.7 V vs. ECS with EDG catalyzed with 20% vanadium oxide, the results showed that the ethylene oxidation process using EDG is efficient. SHEVERDENKIN and colleagues achieved 15 mg L^{-1} of ethanol in a high-pressure reactor with maximum temperatures of 550 °C. The authors also report that ethanol formation reached approximately 5 g of ethanol in 1000 L of pressurized ethane, this efficiency value is lower than that presented by EDG catalyzed with 20% vanadium oxide at a potential of 1.7 V vs. ECS, reaching approximately 82 g of ethanol in 1000 L of ethylene (Sheverdenkin et al, 2004).

The variation in the concentration of ethylene glycol and ethanol using EDG catalyzed with 1% vanadium oxide can best be seen in Figure 76.

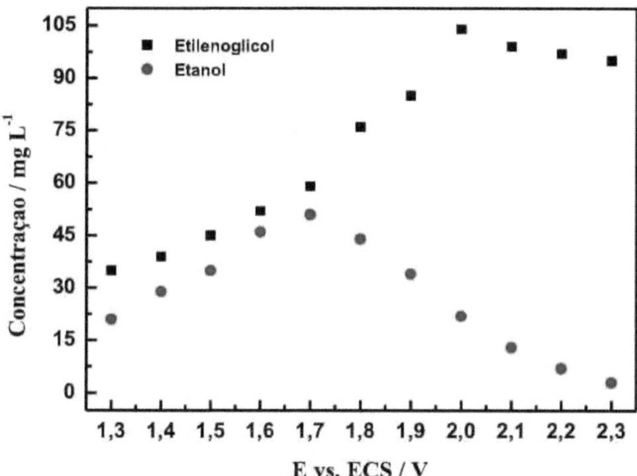

Figura 76 - Variation of the final concentration of ethylene glycol and ethanol as a function of the applied potential using EDG with 1% vanadium

Figure 76 compares the variation in the final concentration of ethylene glycol and ethanol as a function of the potential applied. It can be seen that increasing the potential applied led to an increase in the final concentrations of the products, but ethylene glycol reached higher concentrations, 104 mg L^{-1} at 2.0 V vs. ECS, while ethanol reached its maximum concentration at 1.7 V vs. ECS with 51 mg L^{-1} . At more positive potentials, there was a decrease in the final concentration of the two products, possibly associated with the better potential range for ethylene oxidation for each of the products.

In the experiments catalyzed with 1% vanadium oxide, it was observed that ethylene glycol showed higher concentrations than ethanol at all the potentials applied; this relationship was not observed in the other experiments with different amounts of vanadium. The ratio of the final concentrations of the experiments with 1% vanadium oxide showed characteristics close to those of the experiments with EDG without catalyst (Figure 56), which may be related to the minimum amount of catalyst studied in EDG. This characteristic can be seen in the comparison of the ethylene oxidation results: as the amount of vanadium oxide in the EDG structure increased, the amount of ethanol formed increased and the amount of ethylene glycol decreased, regardless of the potential applied.

When comparing the results of the EDG catalyzed with vanadium oxide, it was observed that the gas diffusion electrodes are efficient in oxidizing ethylene gas, but it is necessary to quantify the chemical efficiency in converting the ethylene molecules into ethylene glycol and ethanol.

The study of chemical efficiency in the oxidation of ethylene and the formation of ethylene glycol is important in order to establish the most suitable pressure value, supplying the EDG with just enough gas for a stoichiometric conversion during the experiment, with the main aim of minimizing the consumption of this reagent during the synthesis process. The efficiency results of the EDGs with different amounts of catalyst are shown in Figure 77.

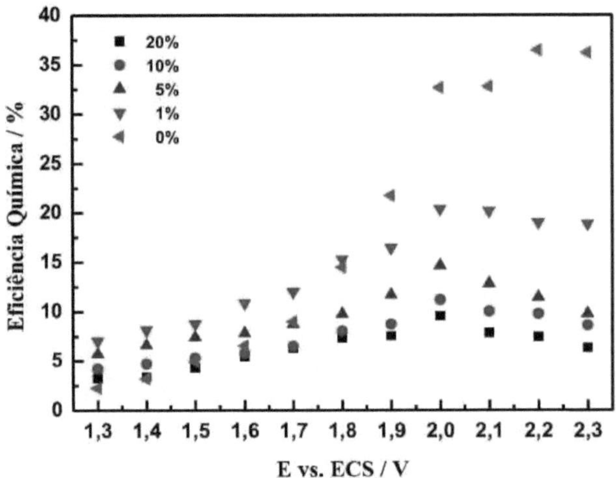

Figura 77 - Variation in the chemical efficiency of ethylene glycol as a function of the applied potential

Figure 77 shows the variation in the chemical efficiency of ethylene glycol formation as a function of the potential applied to the quantities of catalyst studied. There is an increase in the chemical efficiency values with the increase in the applied potential up to the potential of 2.0 V vs. ECS in all the experiments, reaching a minimum efficiency of approximately 9% with 20% vanadium oxide and a maximum efficiency of 20% in the experiments with 1% catalyst. It can be seen in Figure 77 that an increase in the amount of catalyst leads to a decrease in the chemical efficiency of ethylene glycol formation, while an analysis of ethanol formation (Figures 66, 69, 72 and 75) shows that an increase in the amount of catalyst leads to an increase in ethanol formation.

The increase in ethanol formation associated with the decrease in the chemical efficiency of ethylene glycol may be an indication that vanadium oxide interferes with ethylene oxidation, promoting greater ethanol formation to the detriment of ethylene glycol formation, possibly altering the ethylene oxidation mechanism. Looking at Figure 77, we can see the variation in the chemical efficiency values for the formation of ethylene glycol with the increase in the applied potential and with the increase in the amount of catalyst used, but these variations in the efficiency values are not associated with the volume of ethylene used, because the amount of gas used during the experiments was 7.5 mL ±0.4 mL on average, so the variation in chemical efficiency in the formation of ethylene glycol is directly related to the potential applied and the amount of catalyst.

In the study of the oxidation of ethylene to form ethylene glycol using gas diffusion electrodes, it is important to determine the amount of electrical charge applied exclusively to the ethylene glycol formation reaction. To this end, the electrical efficiency was determined and the results are shown in Figure 78.

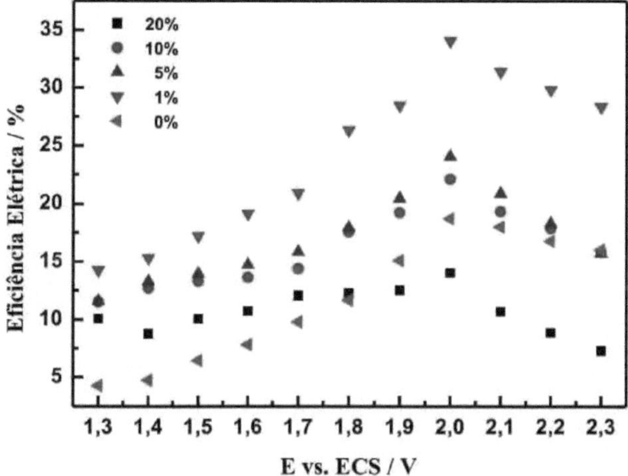

Figura 78 - Variation of electrical efficiency in the formation of ethylene glycol as a function of applied potential

Figure 78 shows the variation in electrical efficiency in the formation of ethylene glycol as a function of the potential applied to the quantities of catalyst studied. It can be seen that increasing the applied potential led to an increase in electrical efficiency up to the potential of 2.0 V vs. ECS, reaching a maximum of 34% when using EDG with 1% vanadium oxide and a minimum of 14% electrical efficiency using EDG with 20% catalyst at 2.0 V vs. ECS. With the results presented, it can be seen that the increase in the amount of catalyst led to a decrease in electrical efficiency in the formation of ethylene glycol. This change, combined with the results for the formation of ethanol (Figures 66, 69, 72 and 75), again shows the tendency of the catalyst to better promote the formation of ethanol from the oxidation of ethylene, reducing the electrical efficiency of the ethylene glycol formation reaction.

In the formation of ethylene glycol by the oxidation of ethylene, it is necessary to determine the amount of energy consumed by the system in the oxidation reaction. To do this, the energy consumption per kilo of ethylene glycol formed was determined and the results are shown in Figure 79.

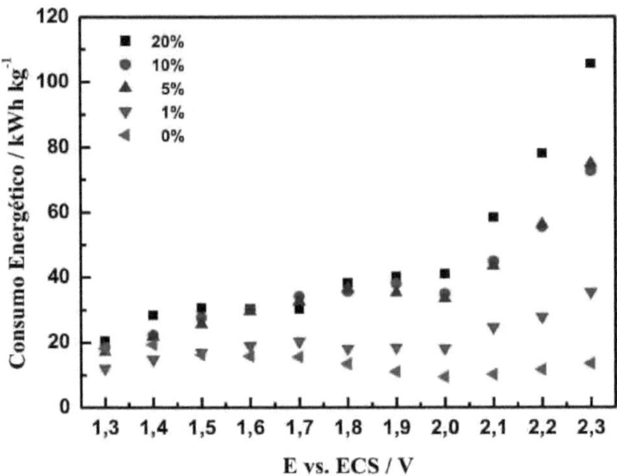

Figura 79 Variation in energy consumption per kilo of ethylene glycol formed as a function of applied potential

Figure 79 shows the variation in the energy consumed in the ethylene glycol formation reaction as a function of the potential applied to the quantities of catalyst studied. It can be seen that the highest amounts of catalyst (20%, 10% and 5%) had energy consumption values close to 1.9 V vs. ECS with approximately 40 kWh kg^{-1} of ethylene glycol, at more positive potentials a greater increase in consumption was observed in the experiment with 20% and in the experiments with 10% and 5% consumption values close to 2.3 V vs. ECS were observed.

It should also be noted that the experiments with 1% vanadium oxide showed the lowest consumption values, possibly due to the lower amount of catalyst, since increasing the amount of vanadium oxide promotes an increase in the amount of ethanol and a consequent decrease in the amount of ethylene glycol.

The results of the experiments with EDG catalyzed with vanadium oxides in different quantities showed a clear tendency for the ethanol formation reaction to be favored over the formation of ethylene glycol. The results also showed that increasing the amount of catalyst led to an increase in the amount of ethanol formed and a decrease in ethylene glycol. The results also showed that increasing the amount of catalyst led to a decrease in the chemical and electrical efficiencies in the formation of ethylene glycol, with a consequent increase in energy consumption for this reaction. With the results presented, the aim was to change the catalyst and use palladium oxide as a new catalyst, in order to achieve better results in the formation of ethylene glycol.

4.5.2 - Ethylene oxidation using TiO2RuO2 EDG catalyzed with palladium

The ethylene gas oxidation results presented showed the ability of TiO2RuO2 EDG to form ethylene glycol and ethanol. The ethylene oxidation process using vanadium oxide-catalyzed EDG showed that the addition of the catalyst led to an increase in the formation of ethanol and a decrease in ethylene glycol. These changes were also observed in the chemical and electrical efficiency, where the efficiency values for the ethylene glycol formation reaction decreased as the amount of catalyst increased.

With the results presented, the aim was to add palladium (in the form of palladium oxide) to the TiO2RuO2 EDG in certain concentrations, in order to study the interference of the addition of the catalyst on the formation of the products and on the efficiency of ethylene glycol oxidation. The results of the experiments catalyzed with palladium oxide are presented below.

- EDG with 20% palladium oxide

In the ethylene oxidation experiments using the EDG of $(TiO_2)_{0.661}(RuO_2)_{0.283}(PdO_2)_{0.056}$, ethylene glycol and ethanol were quantified following the respective calibration curves. The results of ethylene oxidation catalyzed with 20% palladium oxide are presented below.

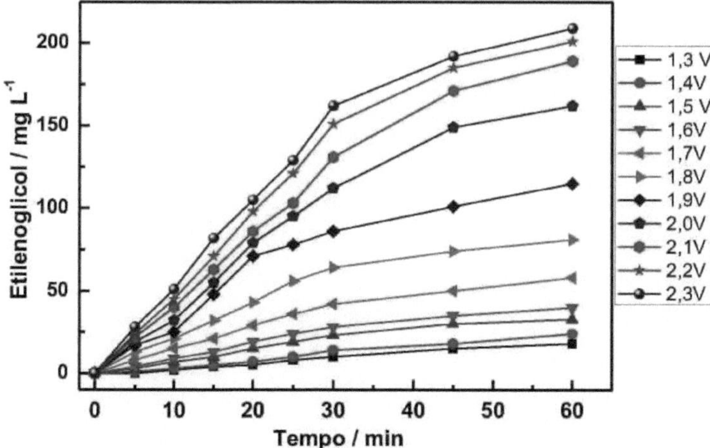

Figura 80 - Variation of ethylene glycol concentration as a function of experiment time using EDG with 20% palladium. Electrolyte: 20 mL Na2SO4 0.1 mol L^{-1}

Figure 80 shows the variation in ethylene glycol concentration as a function of experiment time. The concentration of ethylene glycol increases as the applied potential increases, reaching 18 mg L^{-1} of ethylene glycol at 1.3 V VS. ECS and 209 mg L^{-1} at a potential of 2.3 V VS. ECS. Looking at Figure

80, it can be seen that the concentrations of ethylene glycol show a tendency to stabilize as a function of time from 30 minutes onwards. This stabilization may be associated with the parallel reactions that occur in the oxidation of ethylene.

In the formation of ethylene glycol using EDG catalyzed with palladium oxide, it was observed that the addition of this catalyst promoted an increase in the formation of ethylene glycol, compared to EDG with TiO2RuO2, because EDG without catalyst reached 161 mg L^{-1} and EDG catalyzed with palladium oxide reached 209 mg L^{-1} of ethylene glycol. These results may be associated with an improvement in the ethylene glycol formation reaction due to the action of the catalyst in the ethylene oxidation reaction. One of the parallel reactions to the formation of ethylene glycol is the formation of ethanol. To this end, the ethanol formed during the ethylene oxidation experiments was quantified, and the results of the ethanol formation are shown in Figure 81.

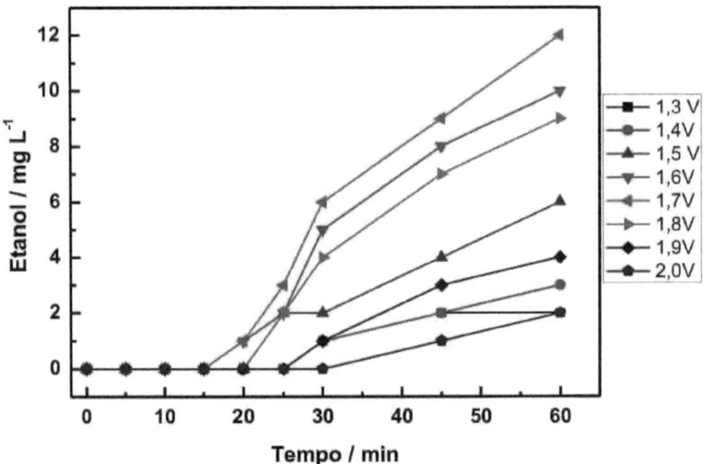

Figura 81 - Variation of ethanol concentration as a function of experiment time using EDG with 20% palladium. Electrolyte: 20 mL Na2SO4 0.1 mol L^{-1}

Figure 81 shows the variation in ethanol concentration as a function of experiment time. It can be seen that the increase in the potential applied led to an increase in the concentration of ethanol, reaching a maximum of 12 mg L^{-1} at the potential of 1.7 V vs. ECS, while at more positive potentials there was a decrease in the concentration of alcohol, reaching a minimum of 2 mg L^{-1} at the potentials of 1.3 V vs. ECS and 2.0 V vs. ECS.

It can also be seen in Figure 81 that ethanol detection only starts after 15 minutes in the experiments at 1.7 V vs. ECS, after 20 minutes in the experiment at 1.8 V vs. ECS and after 30 minutes in the experiment at 2.0 V vs. ECS. ECS, at potentials of 2.1 V, 2.2 V and 2.3 V vs. ECS no ethanol was detected in the samples, possibly because the best potential range for ethanol formation was close to 1.7 V vs. ECS, which is the potential with the best ethanol formation. The variation in the concentration of ethanol and ethylene glycol can be seen in Figure 82.

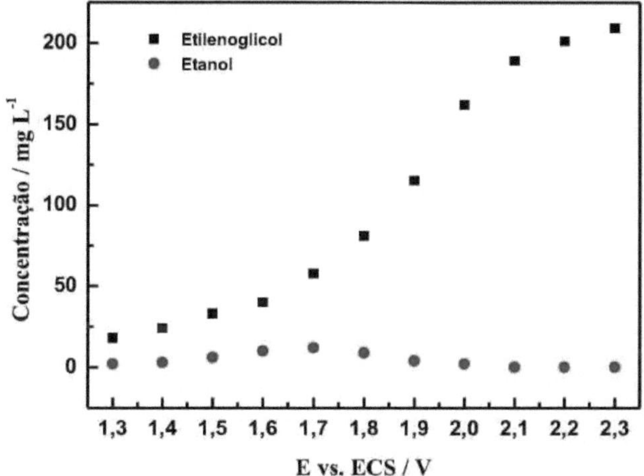

Figura 82 - Variation of the final concentration of ethylene glycol and ethanol as a function of the applied potential using EDG with 20% palladium

Figure 82 shows the variation in the concentration of ethylene glycol and ethanol as a function of the potential applied. It can be seen that the increase in the final concentration of ethylene glycol shows a tendency to stabilize at more positive potentials; this tendency may be associated with the best potential for the formation of ethylene glycol, 2.0 V vs. ECS. With regard to the variation in the final concentrations of ethanol, it can be seen that the maximum concentration is obtained at 1.7 V vs. ECS, at the other potentials concentrations close to zero were obtained, possibly the use of the palladium oxide catalyst may be improving the yield of the ethylene glycol reaction to the detriment of the ethanol reaction.

With regard to the results of the oxidation of ethylene to form ethylene glycol and ethanol, there was an improvement in the results for the formation of ethylene glycol and a decrease in the concentrations of ethanol. Due to the results presented, EDG catalyzed with 10% palladium oxide was used and the results are presented below.

- EDG with 10% palladium oxide

Figures 83 and 84 show the results of the ethylene oxidation experiments using $(TiO_2)_{0.679}(RuO_2)_{0.292}(PdO_2)_{0.029}$ EDG, with 10% palladium oxide.

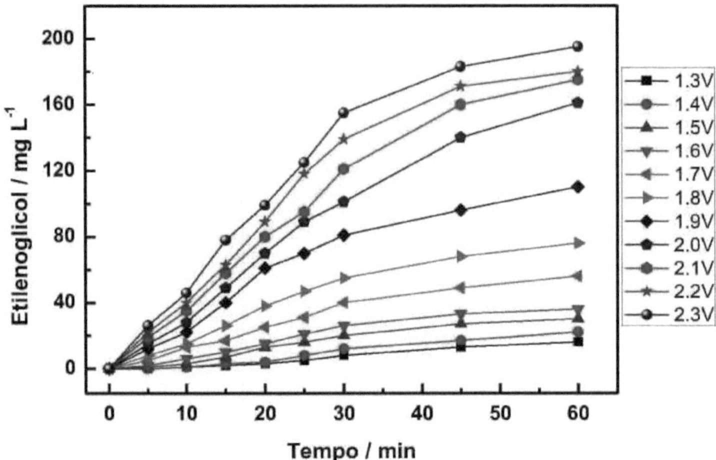

Figura 83 - Variation of ethylene glycol concentration as a function of experiment time using EDG with 10% palladium. Electrolyte: 20 mL Na2SO4 0.1 mol L^{-1}

Figure 83 shows the variation in ethylene glycol concentration as a function of time in experiments using EDG catalyzed with 10% palladium oxide. It can be seen that the concentration of ethylene glycol increases with the increase in potential applied up to 2.3 V vs. ECS, reaching 195 mg L^{-1} , but the concentrations obtained in the experiments with EDG catalyzed with 10% palladium oxide reached lower values compared to the experiments with 20% catalyst, this reduction may be associated with the decrease in the amount of palladium oxide. As with the EDG catalyzed with 20% palladium oxide, Figure 83 shows a stabilization profile from thirty minutes into the experiment. This stabilization profile may be associated with the reactions that occur simultaneously with the reaction to form ethylene glycol from the oxidation of ethylene; an example of these parallel reactions is the reaction to form ethanol, which also originates from the oxidation of ethylene. The results of the formation of ethanol from ethylene are shown in Figure 84.

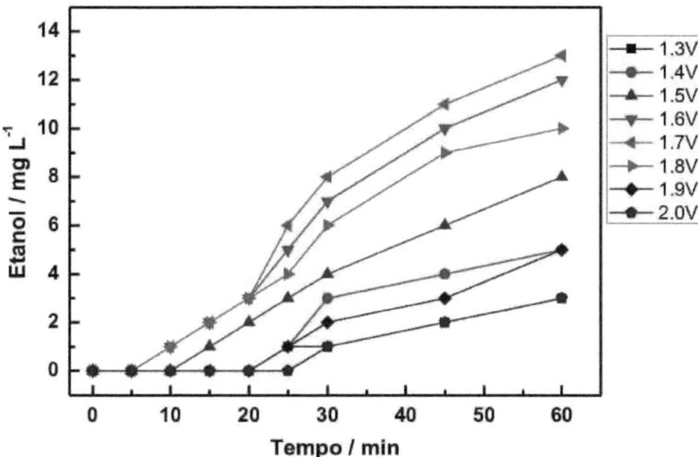

Figura 84 - Variation of ethanol concentration as a function of experiment time using EDG with 10% palladium. Electrolyte: 20 mL Na2SO4 0.1 mol L^{-1}

Figure 84 shows the variation in ethanol concentration as a function of experiment time using EDG catalyzed with 10% palladium. It can be seen that there was an increase in ethanol concentrations in all the experiments, possibly due to the decrease in the amount of catalyst, but the ethanol detection range continued to be between 1.3 V and 2.0 V vs. ECS, remaining the same when compared to the potential range with ethanol detection in the experiments with 20% catalyst.

It can also be seen in Figure 84 that the start of ethanol detection occurred in a shorter period of time compared to the experiments with a greater amount of catalyst, where ethanol was only detected after fifteen minutes of experimentation. With regard to the experiments with 10% catalyst, ethanol detection occurred at 5 minutes of experiment, 1.6 V, 1.7 V and 1.8 V vs. ECS, reaching a maximum concentration of 13 mg L^{-1} of ethanol at 1.7 V vs. ECS and at more positive potentials there was a decrease in ethanol concentration, reaching a minimum of 3 mg L^{-1} at 1.3 V and 2.0 V vs. ECS. The variation in the concentration of ethanol and ethylene glycol is best seen in Figure 85.

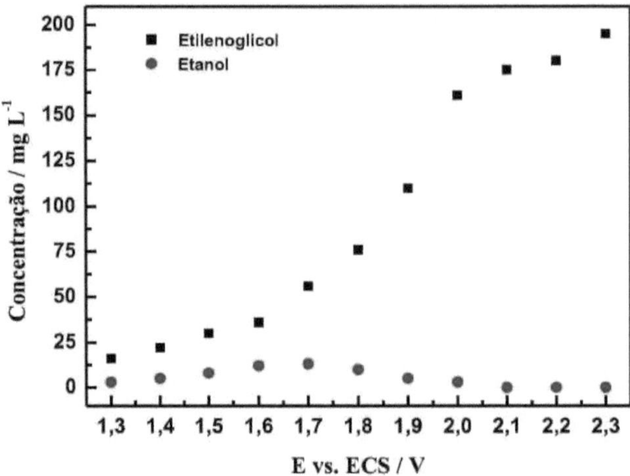

Figura 85 - Variation of the final concentration of ethylene glycol and ethanol as a function of the applied potential using EDG with 10% palladium

Figure 85 shows the variation in the concentration of ethylene glycol and ethanol as a function of the potential applied using gas diffusion electrodes catalyzed with 10% palladium oxide. It can be seen that the increase in the final concentration of ethylene glycol shows a tendency to stabilize at the more positive potentials from 2.0 V vs. ECS, which may be associated with the better potential for the formation of ethylene glycol.

With regard to the variation in the final concentrations of ethanol, it can be seen that the maximum concentration is obtained at 1.7 V vs. ECS, at the other potentials concentrations close to zero were obtained, possibly the use of the palladium oxide catalyst improved the yield of the ethylene glycol reaction with a decrease in the formation of ethanol.

- EDG with 5% palladium oxide

For the ethylene gas oxidation experiments, EDG of $(TiO_2)_{0.689}(RuO_2)_{0.296}(PdO_2)_{0.015}$ with 5% palladium oxide was used and the results are presented below.

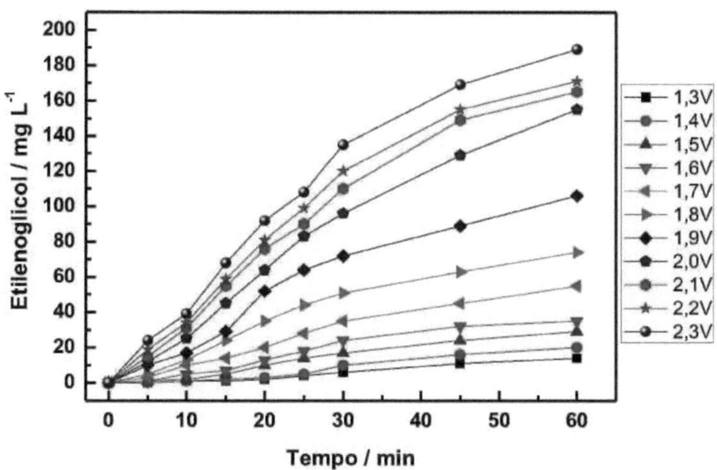

Figura 86 - Variation of ethylene glycol concentration as a function of experiment time using EDG with 5% palladium. Electrolyte: 20 mL Na2SO4 0.1 mol L^{-1}

Figure 86 shows the variation in ethylene glycol concentration as a function of experiment time using EDG catalyzed with 5% palladium oxide. There is an increase in the concentration of ethylene glycol as the applied potential increases up to 2.3 V vs. ECS, reaching 189 mg L^{-1} of ethylene glycol.

Figure 86 shows a decrease in ethylene glycol concentrations in all the experiments compared to the experiments catalyzed with 20% and 10% palladium oxide. This decrease may be associated with a reduction in the amount of catalyst used in each of the electrodes. There was also a tendency for the concentrations to stabilize as a function of time from 30 minutes onwards. This stabilization profile may be associated with the occurrence of parallel reactions to the ethylene glycol formation reaction.

As the amount of catalyst used decreased, the concentrations of ethylene glycol decreased, but an increase in ethanol concentrations was observed in the experiments with a greater amount of catalyst. The results of ethanol formation using EDG catalyzed with 5% palladium oxide are shown in Figure 87.

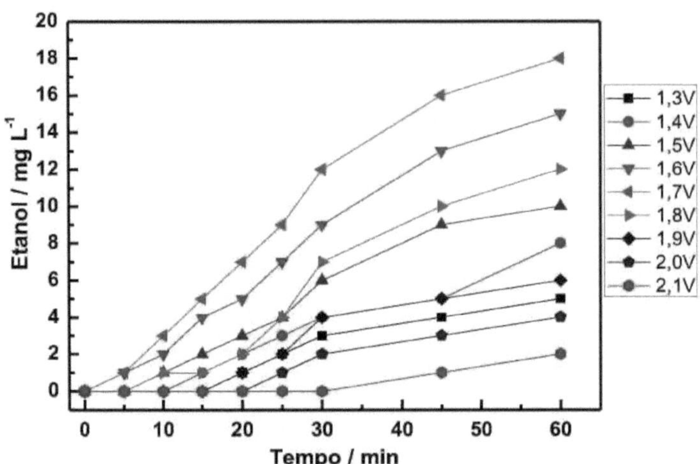

Figura 87 - Variation of ethanol concentration as a function of experiment time using EDG with 5% palladium. Electrolyte: 20 mL Na2SO4 0.1 mol L^{-1}

Figure 87 shows the variation in ethanol concentration as a function of experiment time using gas diffusion electrodes catalyzed with 5% palladium oxide. It can be seen that an increase in the potential applied led to an increase in the concentration of ethanol up to 1.7 V vs. ECS, reaching 18 mg L^{-1} at the end of the one-hour experiment, while at more positive potentials there was a decrease in the concentration of alcohol, reaching a minimum of 2 mg L^{-1} at 2.1 V vs. ECS.

Looking at Figure 87, it can be seen that ethanol was detected in all the samples in the experiment at 1.7 V vs. ECS, unlike the experiments with 20% and 10%, where ethanol was not detected in the first samples. This may be related to the reduction in the amount of catalyst used and the consequent increase in ethanol formation. Another difference observed in these experiments with 5% catalyst was the widening of the potential range where ethanol was detected, with 2.1 V vs. ECS being the most positive potential with ethanol detection.

With the results presented, it can be seen that decreasing the amount of catalyst in the EDG led to a decrease in the concentration of ethylene glycol and an increase in the concentration of ethanol. The variation in the concentration of ethylene glycol and ethanol as a function of the potential applied is shown in Figure 88.

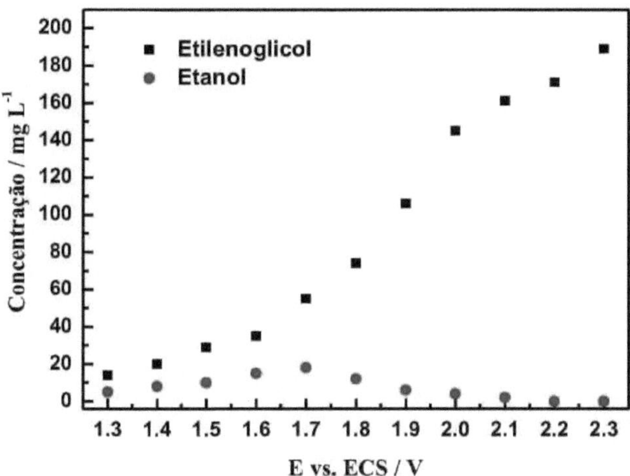

Figura 88 - Variation of the final concentration of ethylene glycol and ethanol as a function of the applied potential using EDG with 5% palladium. Electrolyte: 20 mL $Na2SO4$ 0.1 mol L^{-1}

Figure 88 shows the variation in the concentration of ethylene glycol and ethanol as a function of the potential applied using gas diffusion electrodes catalyzed with 5% palladium oxide. It can be seen that the increase in the final concentration of ethylene glycol shows a tendency to stabilize at the more positive potentials from 2.0 V vs. ECS, this tendency may be associated with the better potential for the formation of ethylene glycol at 2.0 V vs. ECS.

With regard to the variation in the final concentrations of ethanol, it can be seen that the maximum concentration is obtained at 1.7 V vs. ECS, at the other potentials concentrations close to zero were obtained, possibly the use of the palladium oxide catalyst improved the yield of the ethylene glycol reaction with a decrease in the formation of ethanol.

- EDG with 1% palladium oxide

For the ethylene gas oxidation experiments, the $(TiO2)_{0.698}(RuO2)_{0.299}(PdO2)_{0.003}$ EDG was used with the smallest amount of catalyst, 1% palladium oxide, and the results are presented below.

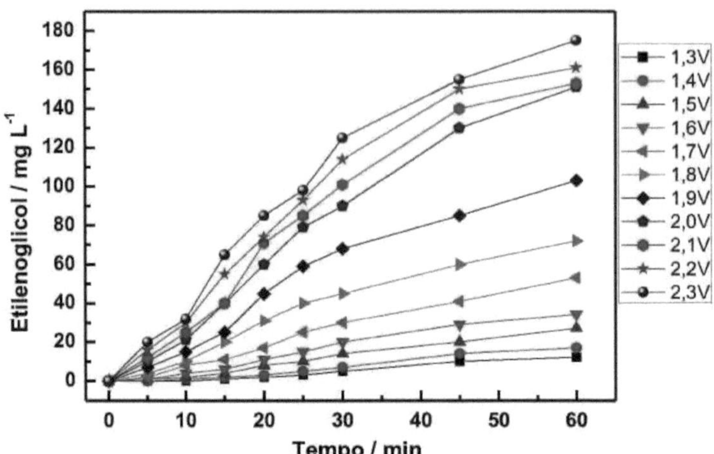

Figura 89 - Variation of ethylene glycol concentration as a function of experiment time using EDG with 1% palladium. Electrolyte: 20 mL Na2SO4 0.1 mol L^{-1}

Figure 89 shows the variation in ethylene glycol concentration as a function of experiment time using the gas diffusion electrode catalyzed with 1% palladium oxide. It can be seen that the increase in the potential applied led to an increase in the formation of ethylene glycol up to the potential of 2.3 V vs. ECS, reaching a maximum concentration of 175 mg L^{-1} at the end of the one-hour experiment. Looking at the figure, we can see a tendency for the concentration of ethylene glycol to stabilize as a function of time after thirty minutes of experimenting. This tendency may be associated with the reactions that occur in parallel with the reaction of ethylene glycol formation from ethylene oxidation.

Comparing the concentrations in Figure 89 with the experiments with the different amounts of catalyst, it can be seen that increasing the amount of palladium oxide led to an increase in the concentration of ethylene glycol at all the potentials studied, reaching 209 mg L^{-1} of ethylene glycol in the experiment at 2.3 V vs. ECS using the gaseous diffusion electrode catalyzed with 20% palladium oxide, with a decrease in the amount of catalyst there was a decrease in ethylene glycol concentrations, reaching 175 mg L^{-1} in the experiment at 2.3 V vs. ECS with 1% catalyst, but these concentrations were higher than the maximum concentration reached by EDG without catalyst, 161 mg L^{-1} of ethylene glycol.

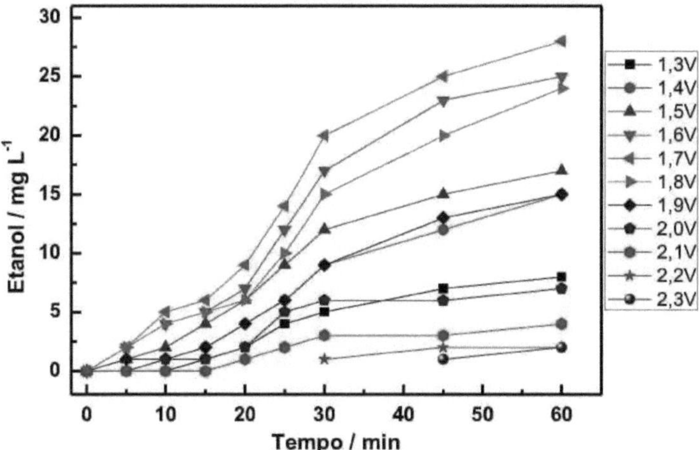

Figura 90 - Variation of ethanol concentration as a function of experiment time using EDG with 1% palladium. Electrolyte: 20 mL Na2SO4 0.1 mol L^{-1}

Figure 90 shows the variation in ethanol concentration as a function of experiment time using EDG with 1% palladium. There is an increase in ethanol concentration as the applied potential increases up to 1.7 V vs. ECS, reaching a maximum concentration of 28 mg L^{-1} at the end of the hour-long experiment. At more positive potentials, the alcohol concentration decreases, reaching a concentration of 2 mg L-1 of ethanol at 2.3 V vs. ECS.

As the catalyst concentration decreased, ethane was detected at all potentials, but at 2.2 V vs. ECS alcohol was only detected after 30 minutes and at 2.3 V vs. ECS alcohol was only detected after 45 minutes. This pattern in ethanol detection may be associated with the action of the catalyst and also with the more positive potentials, which are different from the best potential for ethanol formation, 1.7 V vs. ECS. The variation in the final concentration of ethanol and ethylene glycol as a function of the potential applied can be seen in Figure 91.

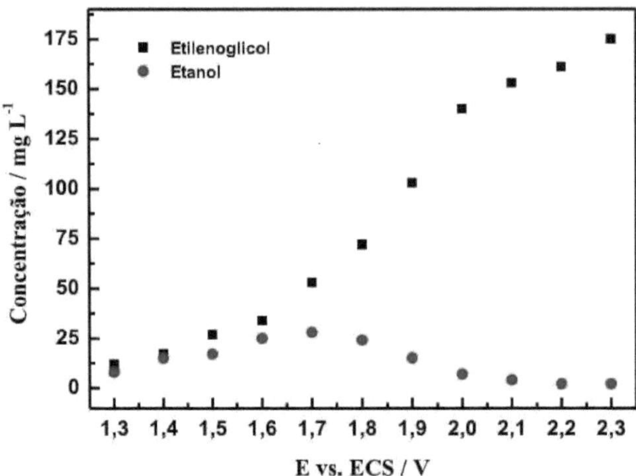

Figura 91 - Variation of the final concentration of ethylene glycol and ethanol as a function of the applied potential using EDG with 1% palladium

Figure 91 shows the variation in the concentration of ethylene glycol and ethanol as a function of the potential applied using gas diffusion electrodes catalyzed with 1% palladium oxide. It can be seen that the increase in the final concentration of ethylene glycol shows a tendency to stabilize at the more positive potentials from 2.0 V vs. ECS, which may be associated with the better potential for the formation of ethylene glycol at 2.0 V vs. ECS. With regard to ethanol, it can be seen that the maximum concentration is obtained at 1.7 V vs. ECS, in the other more positive potentials concentrations close to zero were obtained, possibly the use of the palladium oxide catalyst improved the yield of the ethylene glycol reaction with a decrease in the formation of ethanol, but the decrease in the amount of catalyst promoted an increase in the formation of ethanol at all the potentials applied.

The study of chemical efficiency in the oxidation of ethylene and the formation of ethylene glycol is important in order to establish the most suitable pressure value, supplying the EDG with just enough gas for a stoichiometric conversion during the experiment, with the main aim of minimizing the consumption of this reagent during the synthesis process. The efficiency results of the EDGs with different amounts of catalyst are shown in Figure 92.

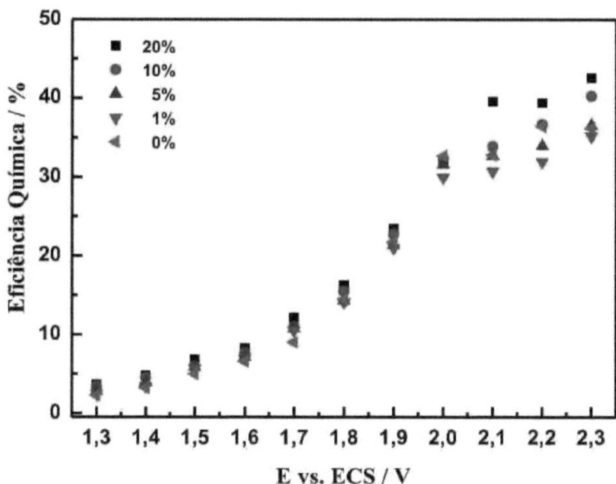

Figura 92 - Variation in the chemical efficiency of ethylene glycol as a function of the applied potential

Figure 92 shows the variation in chemical efficiency in the conversion of ethylene molecules into ethylene glycol as a function of the potential applied. It can be seen that varying the amount of catalyst did not lead to large variations in the chemical efficiency at each potential applied, but increasing the potential applied did lead to an increase in the conversion efficiency of ethylene to ethylene glycol up to the potential of 2.3 V vs. ECS, reaching 42% efficiency with 20% palladium oxide. At the potentials of 2.1 V vs. ECS to 2.3 V vs. ECS, it was observed that the experiments with 20% palladium obtained greater efficiency compared to the experiments with the other amounts of catalyst; this difference may be associated with the better potential range for the formation of ethylene glycol.

The chemical efficiency results showed that varying the amount of catalyst had little effect on the efficiency of converting ethylene into ethylene glycol and that the conversion efficiency was dependent on the potential applied. With regard to the ethylene oxidation reaction, it is necessary to evaluate the efficiency of the electrical charge applied to the ethylene glycol formation reaction. The results of the electrical efficiency are shown in Figure 93.

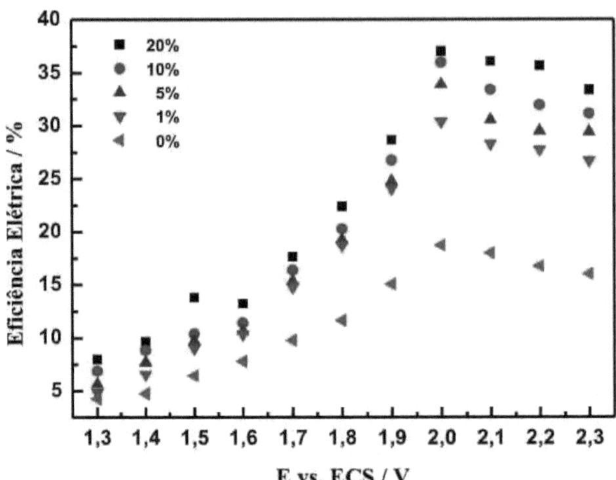

Figura 93 - Variation of electrical efficiency in the formation of ethylene glycol as a function of applied potential

Figure 93 shows the variation in the electrical efficiency values of the ethylene glycol formation reaction as a function of the potential applied. It can be seen that, as with the chemical efficiency, the electrical efficiency values do not change much with the variation in the amount of palladium oxide used in each experiment; it can be seen that the experiments with 20% catalyst showed slightly higher efficiency values than the other experiments.

In the variation of the electrical efficiency values it can be seen that at potentials of 2.1 V, 2.2 V and 2.3 V vs. ECS there was a decrease in electrical efficiency for the ethylene glycol formation reaction, this decrease may be associated with the expenditure of electrical energy for reactions parallel to the ethylene glycol formation reaction, this difference in electrical efficiency at certain applied potentials can be seen in Figure 94.

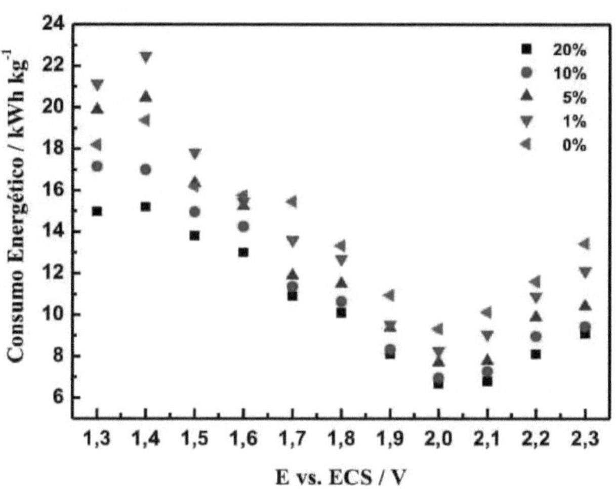

Figura 94 Variation in energy consumption per kilo of ethylene glycol formed as a function of applied potential

Figure 94 shows the variation in energy consumption for the formation of ethylene glycol as a function of the potential applied. It can be seen that energy consumption decreases with increasing applied potential up to 2.0 V vs. ECS, reaching a minimum of approximately 6 kWh kg^{-1} of ethylene glycol using EDG catalyzed with 20% palladium oxide. At more positive potentials, energy consumption increases from 2.0 V vs. ECS, reaching a maximum of 12 kWh kg^{-1} of ethylene glycol using EDG catalyzed with 1% palladium oxide. It can also be seen that, like the chemical and electrical efficiency, the energy consumption values do not change much with the variation in the amount of palladium oxide used in each experiment, but the experiments with 20% catalyst showed slightly lower consumption values than the other experiments.

The variation in energy consumption at more positive potentials is similar to the variation in electrical efficiency in the same potential range. Figure 93 shows a decrease in the electrical efficiency of the ethylene glycol formation reaction between the potentials of 2.0 V and 2.3 V vs. ECS, while in this same potential range there is an increase in energy consumption per kilo of ethylene glycol formed. This decrease in efficiency and increase in energy consumption may be associated with the reactions taking place in parallel with the ethylene glycol formation reaction, thus decreasing the efficiency of applying the electric charge to the main reaction and consequently increasing energy consumption.

With regard to the results presented, the use of palladium oxide promoted a considerable increase in the formation of ethylene glycol and a decrease in the concentration of ethanol, when compared to the EDG without catalyst and the EDG catalyzed with vanadium oxide. It can also be seen that varying the amount of catalyst did not lead to major changes in the values for chemical and electrical efficiency and energy consumption, with the experiments at 2.0 V vs. ECS showing the best efficiency in applying the electrical charge and the lowest energy consumption for the formation of ethylene glycol. With the results of the experiments catalyzed with vanadium oxide and palladium oxides, it was planned to change the catalyst to silver oxide in order to study the changes in the ethylene oxidation reaction.

4.5.3 - Ethylene oxidation using TiO2RuO2 EDG catalyzed with silver

Analysis of the results of ethylene gas oxidation showed the ability of the TiO2RuO2 EDG to form ethylene glycol and ethanol, and the addition of palladium oxide increased the formation of ethylene glycol and decreased the formation of ethanol. With the results presented, the aim was to replace palladium oxide with silver oxide in the TiO2RuO2 EDG at certain concentrations, in order to study the interference of the addition of the catalyst in the formation of the products and in the efficiency of ethylene glycol oxidation. The results of the experiments catalyzed with vanadium are presented below.

- EDG with 20% silver oxide

In the ethylene glycol oxidation experiments using the $(TiO_2)_{0.661}(RuO_2)_{0.283}(AgO)_{0.056}$ EDG, ethylene glycol and ethanol were quantified and the results of the experiments catalyzed with 20% silver oxide are presented below.

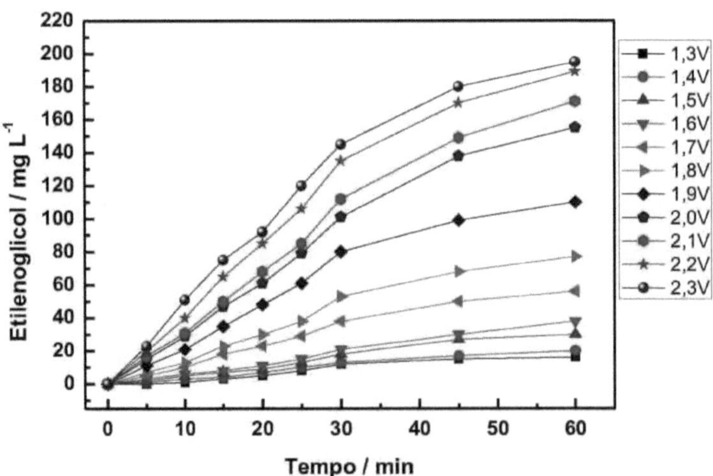

Figura 95 - Variation of ethylene glycol concentration as a function of experiment time using EDG with 20% silver. Electrolyte: 20 mL Na2SO4 0.1 mol L^{-1}

Figure 95 shows the variation in the formation of ethylene glycol as a function of experiment time using EDG catalyzed with 20% silver oxide. It can be seen that the concentration of ethylene glycol increased with the increase in potential applied up to 2.3 V vs. ECS, reaching 195 mg L^{-1} of ethylene glycol at the end of the one-hour experiment. Figure 95 shows a tendency for ethylene glycol concentrations to stabilize as a function of time after 30 minutes of experimenting. This stabilization tendency may be associated with reactions that occur in parallel with the ethylene glycol formation reaction, one of the reactions that occurs in parallel with the ethylene glycol reaction is the reaction to form ethanol from the oxidation of ethylene, the results of the formation of ethanol are shown in Figure 96.

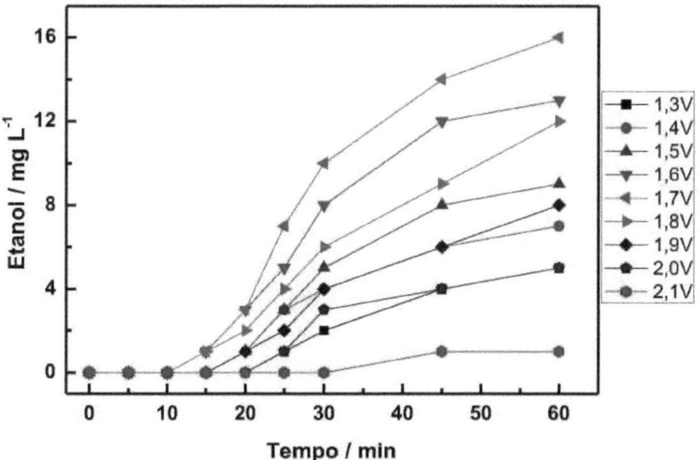

Figura 96 - Variation of ethanol concentration as a function of experiment time using EDG with 20% silver. Electrolyte: 20 mL Na2SO4 0.1 mol L^{-1}

Figure 96 shows the variation in ethanol concentration as a function of experiment time using EDG with 20% silver oxide. It can be seen that increasing the potential applied led to an increase in the concentration of ethanol up to the potential of 1.7 V vs. ECS, reaching 16 mg L^{-1} of alcohol; at more positive potentials there was a decrease in the concentration of ethanol, reaching 2 mg L^{-1} at 2.1 V vs. ECS at the end of the hour-long experiment, and at potentials of 2.2 V and 2.3 V vs. ECS there was no detection of ethanol.

Looking at Figure 96, it can be seen that ethanol is only detected after 15 minutes for the experiments at 1.6 V, 1.7 V and 1.8 V vs. ECS, possibly due to the addition of silver oxide as a catalyst, as the use of vanadium oxide promoted ethanol detection from the start of the experiments. The variation in the concentration of ethanol and ethylene glycol can also be seen in Figure 97.

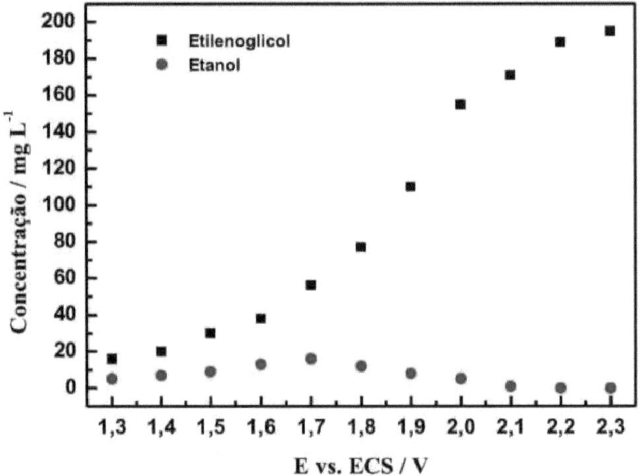

Figura 97 - Variation of the final concentration of ethylene glycol and ethanol as a function of the applied potential using EDG with 20% silver

Figure 97 shows the variation in the final concentration of ethylene glycol and ethanol as a function of the potential applied using EDG with 20% silver oxide. It can be seen that the increase in the applied potential promoted an increase in the formation of ethylene glycol up to the potential of 2.3 V vs. ECS, but a tendency for the concentration to stabilize as a function of the applied potential can be observed, possibly associated with the better potential range for the formation of ethylene glycol. This variation is not observed for the formation of ethanol, since the concentration of alcohol increases with the increase in the applied potential up to 1.7 V vs. ECS, but at more positive potentials a decrease in the final concentration is observed, which is not detected at potentials 2.2 V and 2.3 V vs. ECS, possibly because the best potential for the formation of ethanol is 1.7 V vs. ECS.

The results of the formation of ethanol and ethylene glycol using EDG with 20% catalyst showed that the addition of silver oxide increased the formation of ethylene glycol compared to EDG without catalyst, which is why EDG with 10% silver oxide was used and the results are presented below.

- EDG with 10% silver oxide

In the ethylene glycol oxidation experiments using the $(TiO_2)_{0.679}(RuO_2)_{0.292}(AgO)_{0.029}$ EDG, ethylene glycol and ethanol were quantified and the results of the experiments catalyzed with 10% silver oxide are shown in Figures 98 to 100.

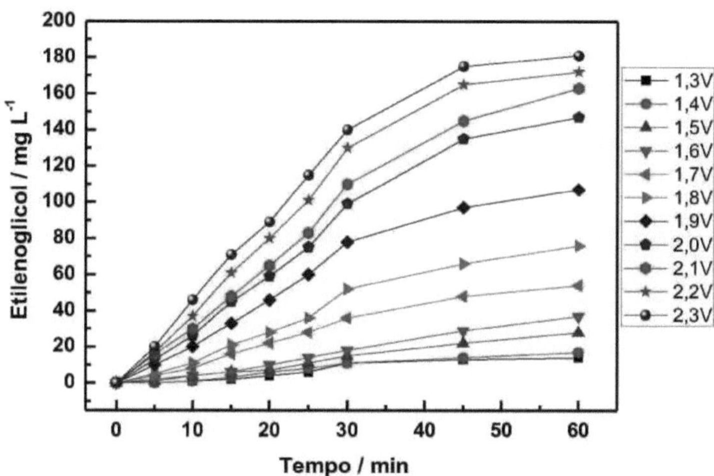

Figura 98 - Variation of ethylene glycol concentration as a function of experiment time using EDG with 10% silver. Electrolyte: 20 mL Na2SO4 0.1 mol L^{-1}

Figure 98 shows the variation in ethylene glycol concentration as a function of experiment time using EDG catalyzed with 10% silver oxide. It can be seen that the concentration of ethylene glycol increases with the increase in the potential applied up to 2.3 V vs. ECS, reaching 181 mg L^{-1} , but the concentrations obtained in the experiments with EDG catalyzed with 10% silver oxide reached lower values compared to the experiments with 20% of the catalyst, this reduction may be associated with the decrease in the amount of silver oxide. As with the other experiments, Figure 98 shows a stabilization profile after thirty minutes of experimenting. This stabilization profile may be associated with the reactions that occur simultaneously with the reaction to form ethylene glycol from the oxidation of ethylene; an example of these parallel reactions is the reaction to form ethanol, which also originates from the oxidation of ethylene. The results of the formation of ethanol from ethylene are shown in Figure 99.

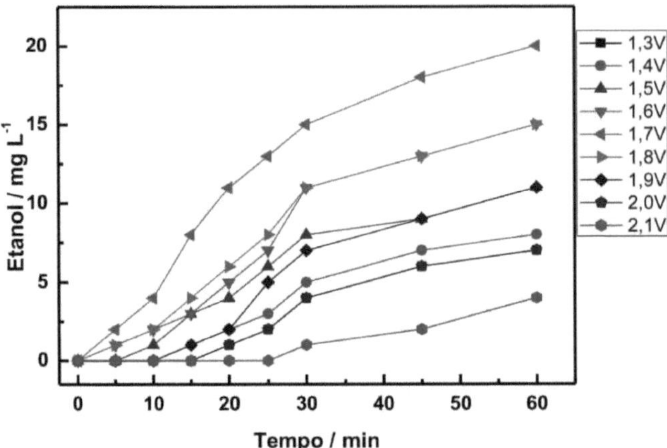

Figura 99 - Variation of ethanol concentration as a function of experiment time using EDG with 10% silver. Electrolyte: 20 mL Na2SO4 0.1 mol L^{-1}

Figure 99 shows the variation in ethanol concentration as a function of experiment time using gas diffusion electrodes catalyzed with 10% silver oxide. It can be seen that an increase in the potential applied led to an increase in the concentration of ethanol up to 1.7 V vs. ECS, reaching 20 mg L^{-1} at the end of the hour-long experiment. At more positive potentials, there was a decrease in the concentration of ethanol, reaching a minimum of 4 mg L^{-1} at 2.1 V vs. ECS.

Looking at Figure 99, it can be seen that ethanol was detected in all the samples in the experiment at 1.7 V vs. ECS. This may be related to the reduction in the amount of catalyst used and the consequent increase in ethanol formation. Another difference observed in these experiments with 10% catalyst was that the potential range in which alcohol was detected remained between 1.3 V and 2.1 V vs. ECS.

With the results presented, it can be seen that decreasing the amount of catalyst in the EDG led to a decrease in the concentration of ethylene glycol and an increase in the concentration of ethanol. The variation in the final concentration of ethylene glycol and ethanol as a function of the applied potential is shown in Figure 100.

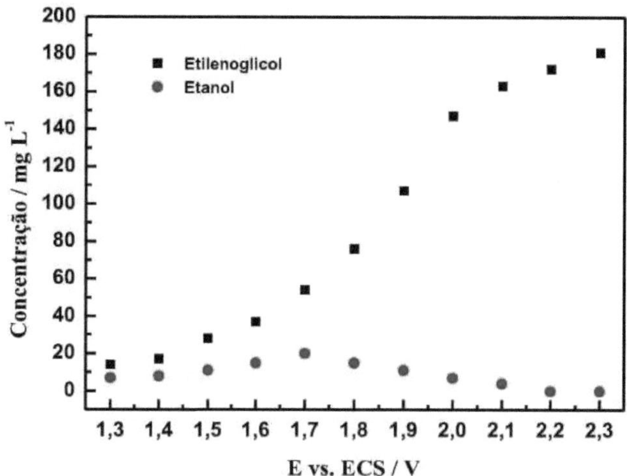

Figura 100 - Variation of the final concentration of ethylene glycol and ethanol as a function of the potential
applied using EDG with 10% silver

Figure 100 shows the variation in the concentration of ethylene glycol and ethanol as a function of the potential applied using gas diffusion electrodes catalyzed with 10% silver oxide. It can be seen that the increase in the final concentration of ethylene glycol shows a tendency to stabilize at the more positive potentials from 2.0 V vs. ECS, which may be associated with the better potential range for the formation of ethylene glycol.

With regard to the variation in the final concentrations of ethanol, it can be seen that the maximum concentration is obtained at 1.7 V vs. ECS, at the other potentials concentrations close to zero were obtained, possibly the use of the silver oxide catalyst improved the yield of the ethylene glycol reaction with a decrease in the formation of ethanol compared to EDG without catalyst.

- EDG with 5% silver oxide

For the ethylene gas oxidation experiments, EDG of $(TiO_2)_{0.689}(RuO_2)_{0.296}(AgO)_{0.015}$ with 5% silver oxide was used and the results are presented below.

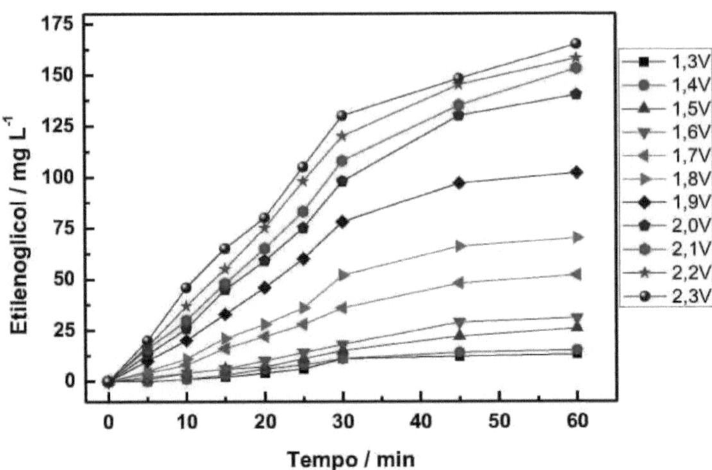

Figura 101 - Variation of ethylene glycol concentration as a function of experiment time using EDG with 5% silver. Electrolyte: 20 mL Na2SO4 0.1 mol L^{-1}

Figure 101 shows the variation in ethylene glycol concentration as a function of experiment time using EDG catalyzed with 10% silver oxide. There is an increase in the concentration of ethylene glycol as the applied potential increases up to 2.3 V vs. ECS, reaching 165 mg L^{-1} of ethylene glycol.

Figure 101 shows a decrease in ethylene glycol concentrations in all the experiments compared to the experiments catalyzed with 20% and 10% silver oxide. This decrease may be associated with a reduction in the amount of catalyst used in each of the electrodes. There was also a tendency for the concentrations to stabilize as a function of time from 30 minutes onwards. This stabilization profile may be associated with the occurrence of parallel reactions to the ethylene glycol formation reaction.

As the amount of catalyst used decreased, the concentration of ethylene glycol decreased, but an increase in the concentration of ethanol was observed in the experiments with 5% catalyst; the results of the formation of ethanol are shown in Figure 102.

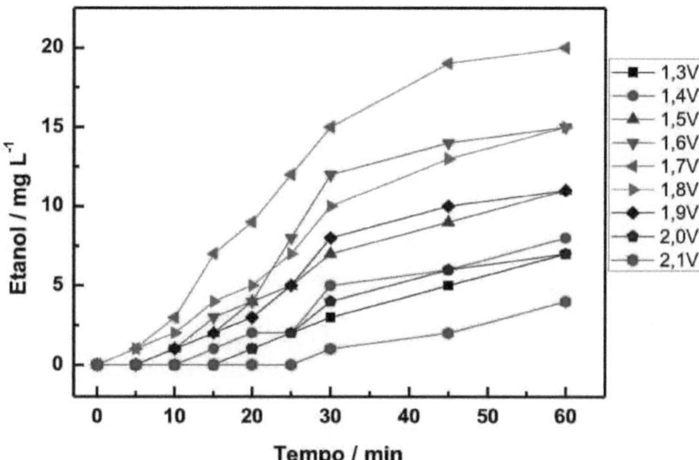

Figura 102 - Variation of ethanol concentration as a function of experiment time using EDG with 5% silver. Electrolyte: 20 mL Na2SO4 0.1 mol L^{-1}

Figure 102 shows the variation in ethanol concentration as a function of experiment time using EDG with 1% silver oxide. There is an increase in ethanol concentration as the applied potential increases up to 1.7 V vs. ECS, reaching a maximum concentration of 20 mg L^{-1} at the end of the hour-long experiment; at more positive potentials there is a decrease in alcohol concentration, reaching a concentration of 4 mg L^{-1} of ethanol at 2.1 V vs. ECS.

As the catalyst concentration decreased, ethane was detected at all the potentials applied, with the exception of 2.2 V and 2.3 V vs. ECS, which, like the previous experiments, obtained alcohol detection between 1.3 V vs. ECS and 2.1 V vs. ECS. The variation in the final concentration of ethanol and ethylene glycol as a function of the potential applied can be seen in Figure 103.

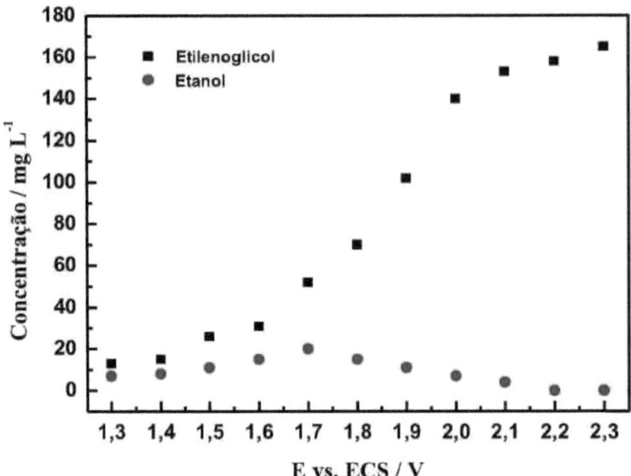

Figura 103 - Variation of the final concentration of ethylene glycol and ethanol as a function of the potential
applied using EDG with 5% silver

Figure 103 shows the variation in the concentration of ethylene glycol and ethanol as a function of the potential applied using gas diffusion electrodes catalyzed with 5% silver oxide. It can be seen that the increase in the final concentration of ethylene glycol shows a tendency to stabilize at the more positive potentials from 2.0 V vs. ECS, which may be associated with the better potential for the formation of ethylene glycol at 2.0 V vs. ECS.

With regard to the variation in the final concentrations of ethanol, it can be seen that the maximum concentration is obtained at 1.7 V vs. ECS, with the final concentration decreasing at the more positive potentials up to 2.1 V vs. ECS.

- EDG with 1% silver oxide

For the ethylene gas oxidation experiments, the $(TiO_2)_{0.698}(RuO_2)_{0.299}(AgO)_{0.003}$ EDG was used with the smallest amount of catalyst, 1% silver oxide, and the results are presented below.

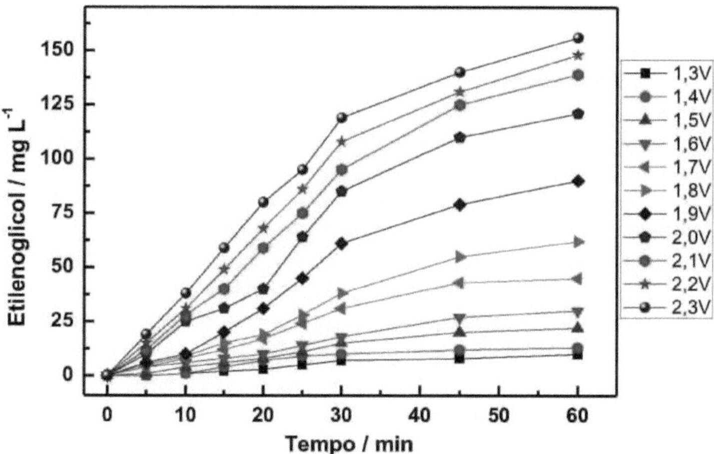

Figura 104 - Variation of ethylene glycol concentration as a function of experiment time using EDG with 1% silver. Electrolyte: 20 mL Na2SO4 0.1 mol L^{-1}

Figure 104 shows the variation in ethylene glycol concentration as a function of experiment time using the gas diffusion electrode catalyzed with 1% palladium oxide. It can be seen that the increase in the potential applied led to an increase in the formation of ethylene glycol up to the potential of 2.3 V vs. ECS, reaching a maximum concentration of 156 mg L^{-1} at the end of the one-hour experiment. Looking at Figure 104, we can see a tendency for the concentration of ethylene glycol to stabilize as a function of time after thirty minutes of experimenting; this tendency may be associated with the reactions taking place in parallel with the ethylene glycol formation reaction.

Comparing the results with the different amounts of catalyst, it can be seen that increasing the amount of silver oxide led to an increase in the concentration of ethylene glycol at all the potentials studied, reaching 195 mg L^{-1} of ethylene glycol in the experiment at 2.3 V vs. ECS using the gas diffusion electrode catalyzed with 20% silver oxide. As the amount of catalyst decreased, the concentrations of ethylene glycol decreased, reaching 156 mg L^{-1} in the experiment at 2.3 V vs. ECS with 1% catalyst, but these concentrations were higher than the maximum concentration reached by the EDG without catalyst, 161 mg L^{-1} of ethylene glycol. The results of ethanol formation as a function of experiment time are shown in Figure 105.

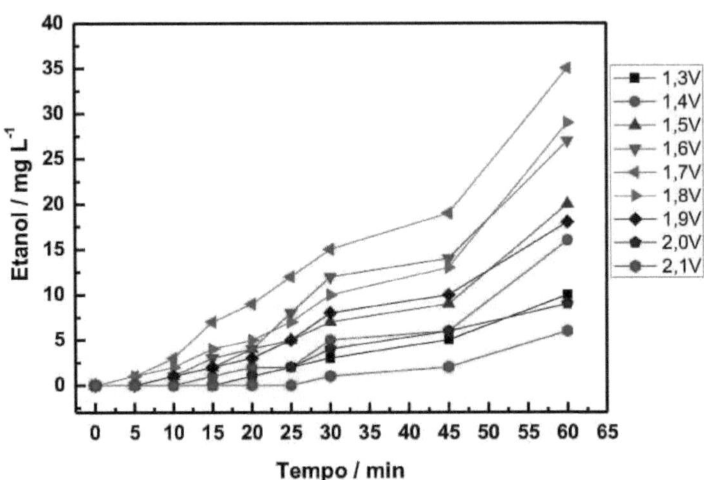

Figura 105 - Variation of ethanol concentration as a function of experiment time using EDG with 1% silver. Electrolyte: 20 mL Na2SO4 0.1 mol L^{-1}

Figure 105 shows the variation in ethanol concentration as a function of experiment time using EDG with 1% silver oxide. There is an increase in ethanol concentration as the applied potential increases up to 1.7 V vs. ECS, reaching a maximum concentration of 35 mg L^{-1} at the end of the hour-long experiment; at more positive potentials there is a decrease in alcohol concentration, reaching a concentration of 6 mg L^{-1} of ethanol at 2.1 V vs. ECS at the end of the hour-long experiment.

Reducing the amount of silver oxide led to an increase in ethanol concentrations, but alcohol was detected between potentials 1.3 V and 2.1 V vs. ECS, the same potential range as the other experiments with different amounts of silver oxide. The variation in the final concentration of ethanol and ethylene glycol as a function of the potential applied can be seen in Figure 106.

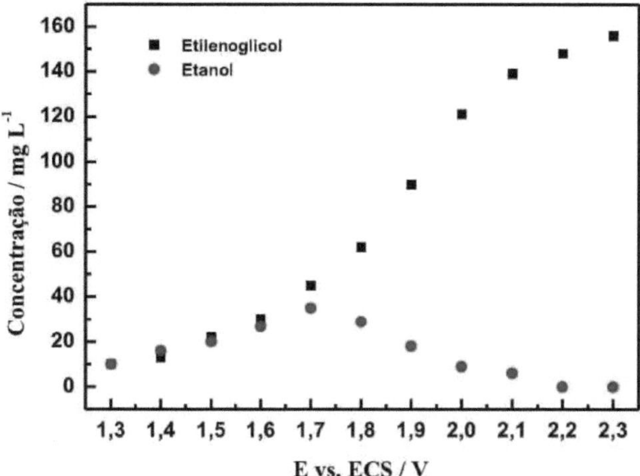

Figura 106 - Variation of the final concentration of ethylene glycol and ethanol as a function of the applied potential
using EDG with 1% silver

Figure 106 compares the variation in the final concentration of ethylene glycol and ethanol as a function of the potential applied. It can be seen that increasing the potential applied led to an increase in the final concentrations of the products, but ethylene glycol reached higher concentrations, 156 mg L^{-1} at 2.3 V vs. ECS, while ethanol reached its maximum concentration at 1.7 V vs. ECS with 35 mg L^{-1} . At more positive potentials, there was a decrease in the final concentration of the two products, possibly associated with the better potential range for ethylene oxidation for each of the products.

In the experiments catalyzed with 1% silver oxide, it was observed that ethylene glycol showed concentrations close to those of ethanol up to the potential of 1.7 V vs. ECS, at more positive potentials, the final concentrations of ethanol decreased while the concentrations of ethylene glycol increased. The ratio of the final concentrations of the experiments with 1% silver oxide showed characteristics close to those of the experiments with EDG without catalyst (Figure 56), which may be related to the minimum amount of catalyst studied in EDG.

In comparing the results of the EDG catalyzed with silver oxide, it was observed that the gas diffusion electrodes are efficient in oxidizing ethylene gas, but it is necessary to quantify the chemical efficiency in converting the ethylene molecules into ethylene glycol and ethanol.

The study of chemical efficiency in the oxidation of ethylene and the formation of ethylene glycol is important in order to establish the most suitable pressure value, supplying the EDG with just enough gas for a stoichiometric conversion during the experiment, with the main aim of minimizing the consumption of this reagent during the synthesis process. The efficiency results of the EDGs with different amounts of catalyst are shown in Figure 107.

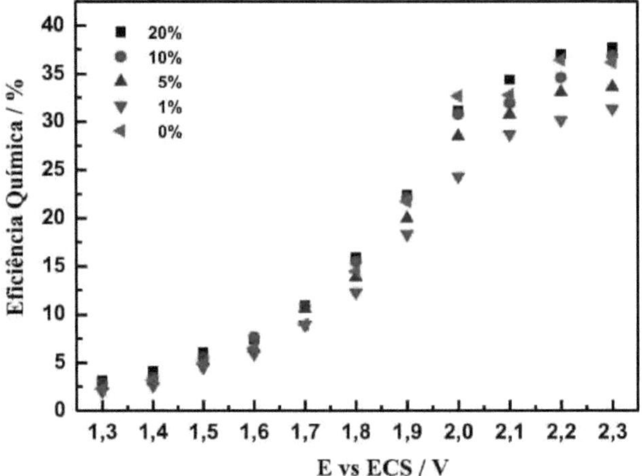

Figura 107 - Variation in the chemical efficiency of ethylene glycol as a function of the applied potential

Figure 107 shows the variation in chemical efficiency in the conversion of ethylene molecules into ethylene glycol as a function of the potential applied. It can be seen that varying the amount of catalyst did not lead to large variations in the chemical efficiency at each potential applied, but increasing the potential applied did lead to an increase in the conversion efficiency of ethylene to ethylene glycol up to the potential of 2.3 V vs. ECS, reaching 37% efficiency with 20% palladium oxide. At potentials of 2.1 V vs. ECS to 2.3 V vs. ECS.

The chemical efficiency results showed that varying the amount of catalyst had little effect on the efficiency of converting ethylene into ethylene glycol and that the conversion efficiency was dependent on the potential applied. With regard to the ethylene oxidation reaction, it is necessary to evaluate the efficiency of the electrical charge applied to the ethylene glycol formation reaction. The results of the electrical efficiency are shown in Figure 108.

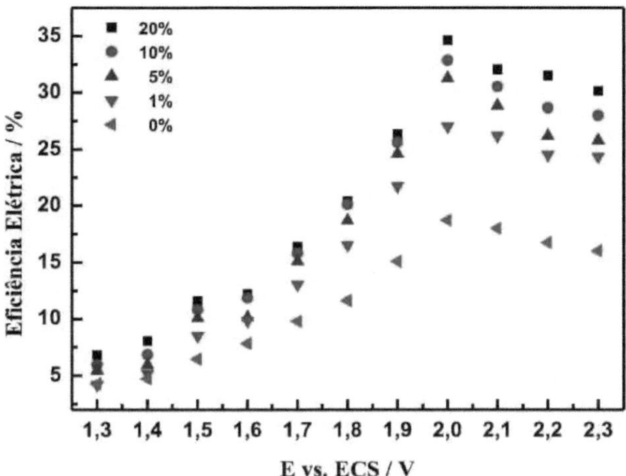

Figura 108 - Variation of electrical efficiency in the formation of ethylene glycol as a function of applied potential

Figure 108 shows the variation in the electrical efficiency values of the ethylene glycol formation reaction as a function of the potential applied. It can be seen that, like the chemical efficiency, the electrical efficiency values do not change much as the amount of silver oxide used in each experiment varies, reaching a maximum of 34% at 2.0 V vs. ECS with 20% catalyst.

In the variation of electrical efficiency values, it can be seen that at potentials of 2.1 V, 2.2 V and 2.3 V vs. ECS there was a decrease in electrical efficiency for the ethylene glycol formation reaction, reaching 30% efficiency; this decrease may be associated with the electrical current applied in the reactions parallel to the ethylene glycol formation reaction; this difference in electrical efficiency at certain applied potentials can be seen in Figure 109.

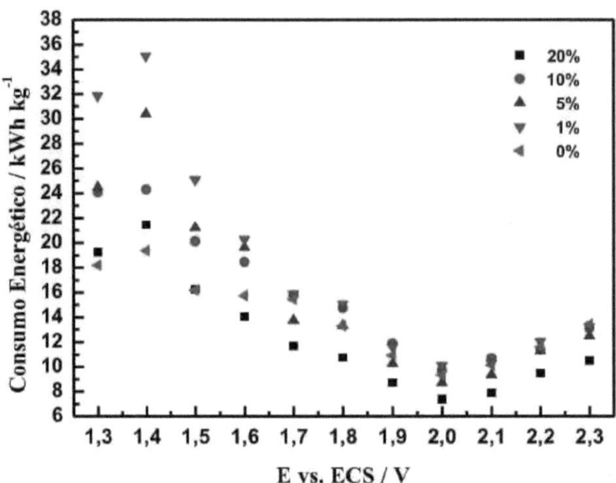

Figura 109 - Variation in energy consumption per kilo of ethylene glycol formed as a function of applied potential

Figure 109 shows the variation in energy consumption for the formation of ethylene glycol as a function of the potential applied. It can be seen that energy consumption decreases with increasing applied potential up to 2.0 V vs. ECS, reaching a minimum of approximately 7 kWh kg^{-1} of ethylene glycol using EDG catalyzed with 20% palladium oxide. At more positive potentials, energy consumption increases from 2.0 V vs. ECS, reaching a maximum of 13 kWh kg^{-1} of ethylene glycol using EDG catalyzed with 1% palladium oxide at a potential of 2.3 V vs. ECS.

The variation in energy consumption at more positive potentials is similar to the variation in electrical efficiency in the same potential range. Figure 108 shows a decrease in the electrical efficiency of the ethylene glycol formation reaction between the potentials of 2.0 V and 2.3 V vs. ECS, while in this same potential range there is an increase in energy consumption per kilo of ethylene glycol formed. This decrease in efficiency and increase in energy consumption may be associated with the reactions taking place in parallel with the ethylene glycol formation reaction, thus decreasing the efficiency of applying the electrical charge to the main reaction and consequently increasing energy consumption.

4.5.4 - Ethylene oxidation using constant current experiments

The results presented show that EDG is efficient in the formation of ethylene glycol from the oxidation of ethylene, reaching a maximum of 161 mg L^{-1} at 2.3 V vs. ECS in the experiments with electrodes without catalyst. With the addition of catalysts, an increase in ethylene glycol formation was observed when palladium oxide was used as a catalyst, reaching 209 mg L^{-1} in the experiment at 2.3 V vs. ECS and 20% catalyst.

The results showed that the maximum ethylene glycol detected was at a potential of 2.3 V vs. ECS, in the experiments without catalyst and with 20% palladium oxide, but the efficiency in the application of the electrical charge in the ethylene glycol formation reaction showed that the most efficient potential was 2.0 V vs. ECS. This difference between the potential with the highest ethylene glycol production and the potential with the highest electrical efficiency may be associated with the best potential range for the ethylene glycol formation reaction, since in all the experiments there was a tendency for the final concentrations to stabilize as a function of the potential applied from 2.0 V vs. ECS, possibly as a result of interference from reactions occurring in parallel with the ethylene glycol formation reaction.

Due to the difference in the potencies of higher production and higher efficiency for ethylene glycol, constant current experiments were used for the formation of ethylene glycol using EDG with 20% palladium oxide, due to the better yield of this catalyst in the formation of ethylene glycol. The results of the formation of ethylene glycol in constant current are presented below.

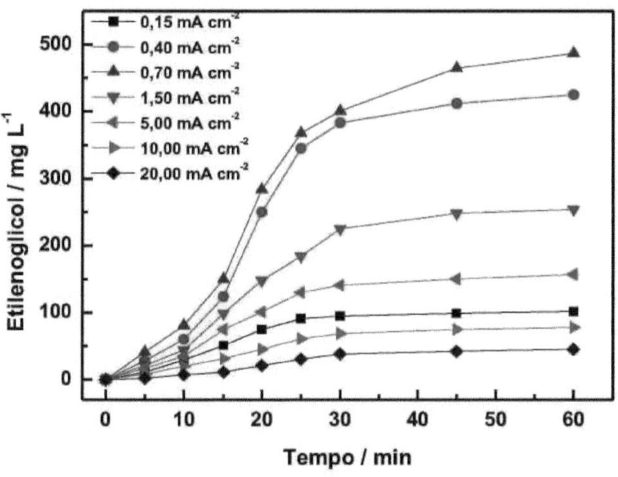

Figura 110 - Variation of ethylene glycol concentration as a function of electrolysis time. Electrolyte: 20 mL Na2SO4 0.1 mol L^{-1}

Figure 110 shows the variation in ethylene glycol concentration as a function of experimental time, using EDG catalyzed with 20% palladium. In all the experiments, there is a tendency for the concentration of ethylene glycol to stabilize as a function of experimental time, possibly due to the reactions that occur in parallel with the ethylene glycol formation reaction.

Figure 110 shows an increase in the concentration of ethylene glycol with the increase in the current density applied up to 0.7 mA cm^{-2} reaching 487 mg L^{-1} of ethylene glycol at the end of an hour's experiment, at more positive potentials it was observed that the concentration decreased with the increase in the current density applied, reaching 45 mg L^{-1} at 20.00 mA cm .$^{-2}$

Looking at the results in Figure 110, it can be seen that the highest concentration achieved was 487 mg L^{-1} at 0.7 mA cm^{-2} , this result was superior to the experiments without catalyst, where the best result for ethylene glycol formation was 341 mg L^{-1} at the same current density, showing that the addition of palladium oxide as a catalyst improved the ethylene glycol formation process. The variation in the final concentration of ethylene glycol in the constant current experiments can be seen in Figure 111.

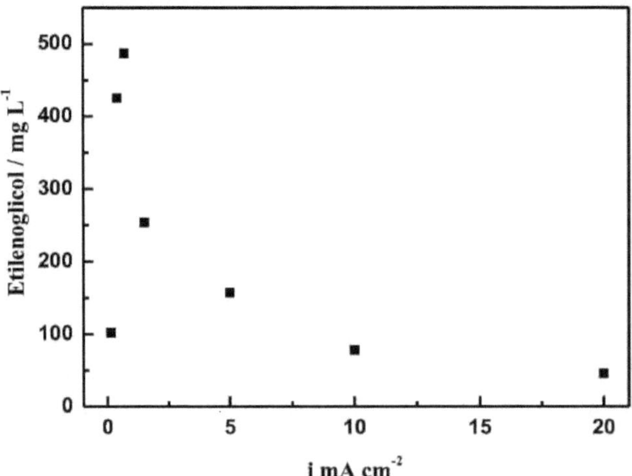

Figura 111 - Variation of the final concentration of ethylene glycol as a function of the current density applied. Electrolyte: 20 mL Na2SO4 0.1 mol L^{-1}

Figure 61 shows the variation in the final concentration of ethylene glycol as a function of the current density applied using EDG catalyzed with 20% palladium oxide. The final concentration of ethylene glycol increased with the current applied, starting at 0.15 mA cm^{-2} with 102 mg L^{-1} of ethylene glycol and reaching a maximum of 487 mg L^{-1} of ethylene glycol with a current of 0.7 mA cm^{-2} , at higher currents, the concentration of ethylene glycol decreased with increasing current density, reaching the lowest value at 20.00 mA cm^{-2} with a concentration of 45 mg L^{-1} of ethylene glycol.

This decrease in ethylene glycol concentration at currents above 0.7 mA cm^{-2} may be associated with the excess current density applied to the ratio of gas volume to active area of the EDG, so the excess current density could be generating other reactions, parallel to the formation of ethylene glycol, such as the O_2 release reaction. The decrease in ethylene glycol formation with increasing current density can also be seen in the variation in chemical efficiency, Figure 112.

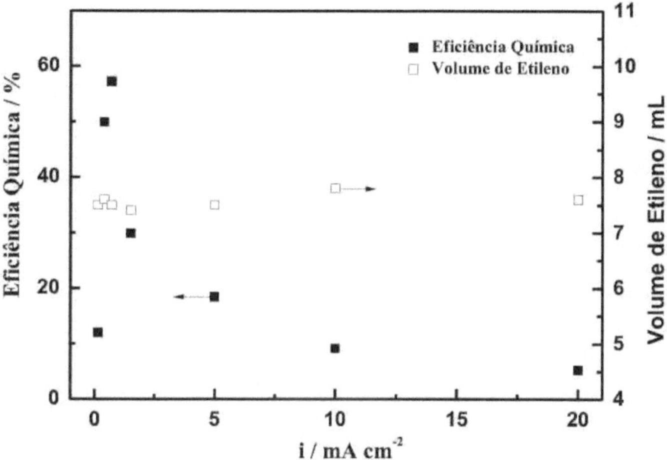

Figura 112 - Variation in chemical efficiency (%) and the volume of ethylene supplied (mL) both as a function of the current density applied.

Figure 112 shows the variation of the chemical efficiency values as a function of the applied potential using gas diffusion electrodes catalyzed with 20% palladium oxide. It can be seen that the chemical efficiency values reached a maximum of approximately 57% at a current density of 0.7 mA cm^{-2} , at higher current densities there was a decrease in the efficiency values, reaching a minimum of 5% at 20.00 mA cm^{-2} , these results are higher than those obtained with EDG without catalyst, where they reached a chemical efficiency of 51% for the formation of ethylene glycol from the oxidation of ethylene.

It can also be seen that the increase in chemical efficiency is associated with the increase in the current density applied and is not related to the variation in the volume of gas during the experiments, because the volume of ethylene supplied varied slightly by around 8 mL per hour of experiment. In the constant current ethylene oxidation reaction with the formation of ethylene glycol, it is important to establish the efficiency of the application of the electrical charge in the ethylene glycol formation reaction; the electrical efficiency results are shown in Figure 113.

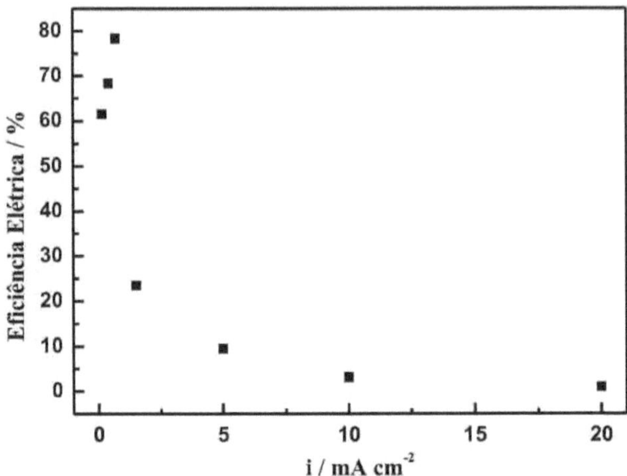

Figura 113 - Variation in Electrical Efficiency (%) as a function of applied current

Figure 113 shows the variation in electrical efficiency as a function of the current density applied using gas diffusion electrodes with 20% palladium oxide. It can be seen that the efficiency increases with the increase in the current density applied up to 0.7 mA cm^{-2} , reaching an approximate efficiency of 78%. At higher current densities there was a decrease in electrical efficiency, reaching a minimum value of 0.9% at 20.00 mA cm^{-2} , these results are higher than those obtained with EDG without catalyst, 66% maximum efficiency for the formation of ethylene glycol from the oxidation of ethylene. This decrease may be associated with the better current density of 0.7 mA cm^{-2} for ethylene oxidation and at higher current densities the formation of ethylene glycol decreases, thus imposing a rapid decrease in efficiency at higher current densities.

The electrical efficiency results show that charge transfer for the oxidation reaction of ethylene to ethylene glycol is more efficient than the chemical transformation of the ethylene molecule into ethylene glycol, even with a charge transfer of approximately 78% (0.7 mA cm^{-2}) the chemical conversion was 57%, indicating that even applying 78% of the total charge of the experiment, the EDG achieved conversion of approximately 57% of the ethylene molecules into ethylene glycol.

With regard to the application of the electric charge, it is important to study the energy consumption involved in the formation of ethylene glycol, Figure 114 shows the energy consumption as a function of the current density applied.

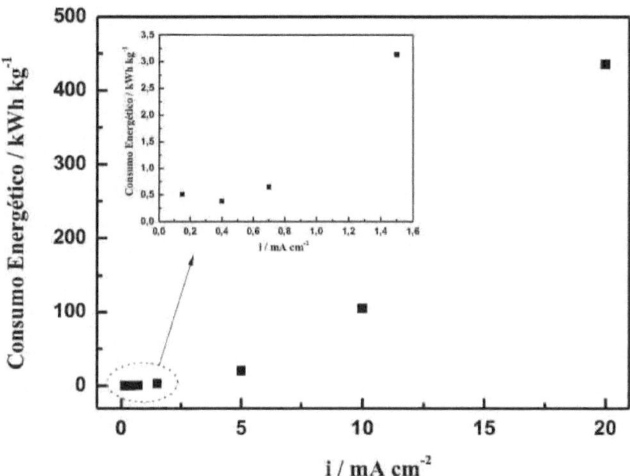

Figura 114 - Variation in energy consumption of the ethylene glycol formation reaction as a function of applied current density

Figure 114 shows the variation in the energy consumed in the ethylene glycol formation reaction as a function of the current applied. It can be seen that energy consumption increases as the current density applied increases, with a greater increase in consumption from 1.5 mA cm^{-2} as shown in Figure 114, reaching a maximum value of approximately 780 kWh kg^{-1} of ethylene glycol at 20 mA cm^{-2} . Figure 114 shows the variation in energy consumption up to a current density of 1.5 mA cm^{-2} , showing that the lowest energy consumption was at 0.4 mA cm^{-2} with 0.4 kWh kg^{-1} followed by the currents 0.15 mA cm^{-2} and 0.7 mA cm^{-2} both with 0.5 kWh kg^{-1} of ethylene glycol formed in one hour of experimentation.

The results shown in Figure 114 are lower than those obtained in the experiments with EDG without catalyst, where they reached 0.5 kWh kg^{-1} at 0.4 mA cm^{-2} and 0.9 kWh kg^{-1} at 0.7 mA cm^{-2} , these results showed that the addition of the palladium oxide catalyst reduced the energy consumption of the ethylene glycol formation reaction.

4.5.5 - Tests on the stability of EDG in the formation of ethylene glycol

The results presented showed that the EDG catalyzed with 20% palladium produced the highest concentration of ethylene glycol, but it is necessary to know the stability of the catalyzed EDG for the formation of ethylene glycol, maintaining its physical and electrochemical characteristics. For this study, three consecutive tests were carried out with a new EDG to determine the reproducibility of the electrode in the ethylene glycol formation process using the parameters of the experiment that produced the highest concentration of ethylene glycol, 2.3 V vs. ECS with 20% palladium oxide.

Before the tests began, the EDG was analyzed using EDX and SEM to quantify the metals on its surface. This analysis was carried out again on the same electrode at the end of the third test and the results compared before and after the experiments are shown in Table 5.

Table 5 - Pre-test and post-test quantities (%) of vanadium-catalyzed EDG

	Titanium		**Ruthenium**		**Palladium**	
	Pre-testing	**Post-tests**	**Pre-testing**	**Post-tests**	**Pre-testing**	**Post-tests**
20% V_2O_5	66,2	66,5	28,3	28,1	5,5	5,4

Table 5 shows the amounts of metals on the surface of EDG catalyzed with 20% palladium before and after the 3 reproducibility tests. There was a slight decrease in the amount of ruthenium (28.3% to 28.1%) and palladium (5.5% to 5.4%) after the reproducibility tests in the formation of ethylene glycol at a potential of 2.3 V vs. ECS. This decrease in the amount of ruthenium and palladium may be associated with reactions on the EDG surface, probably releasing these metals into the electrolyte.

With regard to the tests to reproduce the formation of ethylene glycol, the potential of 2.3 V vs. ECS was chosen because this potential was the best for the formation of ethylene glycol, achieving the highest concentrations in all the tests catalyzed with palladium oxide. The results of the 3 experiments are shown in Figure 115.

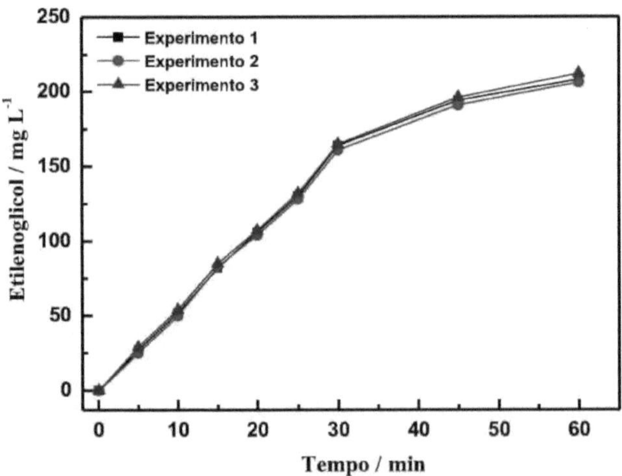

Figura 115 - Variation of ethylene glycol concentration, in triplicate, as a function of experiment time

Figure 115 shows the triplicate tests of the variation in ethylene glycol concentration as a function of time using EDG catalyzed with 20% palladium at a potential of 2.3 V vs. ECS. It can be seen that the concentrations of ethylene glycol were close in all three experiments. This proximity in the concentrations is associated with the reproducibility of the gas diffusion electrode in the process of ethylene glycol formation. It should also be noted that the variation in ethylene glycol concentrations was similar to that observed in previous experiments, with a tendency for concentrations to stabilize as a function of time from 30 minutes onwards. In the comparison with the experiment at 2.3 V vs. ECS with 20% catalyst in Figure 80, it can be seen that the ethylene glycol concentration values are similar when compared to the repetition of the experiments under the same conditions, Figure 115, reaching final concentrations of 208 mg L^{-1} for experiment 1, 206 mg L^{-1} for experiment 2 and 211 mg L^{-1} for experiment 3, compared to 209 mg L^{-1} for the previous experiment (Figure 80). With the results presented, it can be seen that the EDG catalyzed with palladium oxide shows considerable reproducibility for the results of the formation of ethylene glycol from the oxidation of ethylene. Looking at the results shown in Figure 115, we can see a tendency for the concentration of ethylene glycol to stabilize as a function of the experiment time. To better define the trend of stabilization of ethylene glycol concentrations, a 4-hour experiment was carried out with a new EDG, under the experimental conditions of greatest ethylene glycol formation, 2.3 V vs. ECS with EDG catalyzed with 20% palladium oxide and the results are shown in Figure 116.

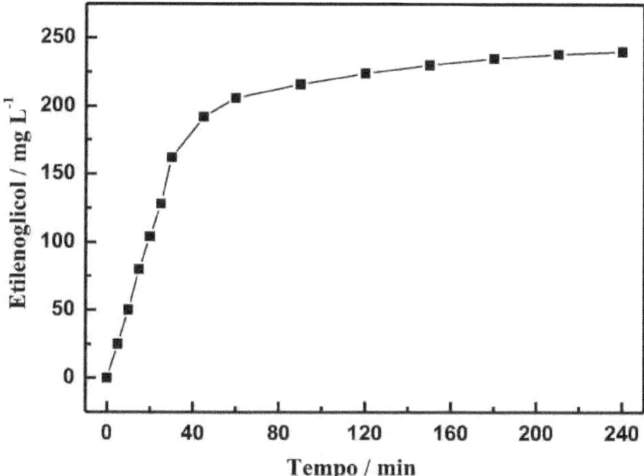

Figura 116 - Variation of ethylene glycol concentration as a function of experiment time

Figure 116 shows the variation in ethylene glycol concentration as a function of time in an experiment lasting 4 hours, using a gas diffusion electrode catalyzed with 20% palladium. It is clear that the concentration of ethylene glycol stabilizes as a function of time from 30 minutes onwards, as observed in previous experiments at different applied potentials and different catalysts. This stabilization may be associated with the reactions that occur in parallel with the reaction to form ethylene glycol from the oxidation of ethylene. It is also observed that there is no decrease in the concentration of ethylene glycol during the experiment, possibly due to the balance between the reactions that occur in parallel from the oxidation of ethylene.

5 Conclusion

The following are the conclusions regarding the steps taken.

5.1 - Preparation/characterization of the metal oxides and construction of the oxide EDG

The catalytic mass used in the construction of the EDG was prepared from the powder of metallic oxides, these oxides had particles mostly between 1 and 4 μm and X-ray analysis showed the formation of Ti/Ru, Ti/Ru/V2O5, Ti/Ru/PdO2 and Ti/Ru/AgO, confirming the efficiency of the calcination process of the polymer precursor solution.

The metal oxide powder was used to prepare the catalytic paste and then the gas diffusion electrode (GDE) made of metal oxides. The EDGs were analyzed in the SEM and the micrographs showed that the electrodes had a compact surface, without cracks or cracks, but the electrodes had some areas with larger particles, which could impose preferential passages for the working gas.

The surface of the EDG was analyzed by EDX to quantify each of the metals. The results showed that the expected theoretical quantities obtained values close to the actual quantities analyzed, but in some cases the actual quantity was different from the theoretical quantity, as in the EDG catalyzed with 1% vanadium, where the difference reached 66% between actual and theoretical. The EDGs catalyzed with 1% silver and 1% palladium had a 33% difference between the theoretical and real values quantified by EDX. The distribution of metals on the surface of the electrodes was also analyzed. It was observed that some electrodes showed agglomerations of certain metals in specific regions on the surface, but in general the metals were detected in all regions of the surfaces analyzed.

The electrochemical characterization tests on the gas diffusion electrodes showed a good balance in the anodic/cathodic charges, where the charge ratio was close to 1.02. With regard to o2 release currents, the electrode showed an increase in current of approximately 1.5 V vs. ECS and with regard to lifetime under working conditions, the EDG showed a potential of 2.0 V vs. ECS for 70 minutes.

5.2 - Methane oxidation in preliminary oxide EDG tests

The gas diffusion electrodes built with Ti/Ru and Ti/Ru/V2O5 in different proportions were previously tested with methane as the working gas and the results showed that the EDG has the characteristic of oxidizing methane to form methanol, formaldehyde and thermal acid.

When EDG was used without a catalyst for methane oxidation, 141 mg L^{-1} of methanol was quantified at the applied potential of 2.2 V vs. ECS, with a chemical efficiency for converting methane into methanol of 5.5%, an electrical efficiency for the methanol formation reaction of 21% with an energy consumption of approximately 11.2 kWh kg^{-1} of methanol formed. Formaldehyde and ferric acid were also detected in the mass spectra.

When EDG catalyzed with vanadium oxide in different proportions was used, it was noted that increasing the amount of catalyst led to an increase in the amount of methanol formed, reaching 340 mg L^{-1} at the end of the one-hour experiment when EDG catalyzed with 20% vanadium oxide was used and formaldehyde was detected at a maximum concentration of 19 mg L^{-1} with the same electrode.

As the amount of vanadium oxide in the EDG decreased, there was a decrease in the maximum concentration of methanol reached at the end of the experiments, with 200 mg L^{-1} of methanol with EDG catalyzed with 5% vanadium. Formaldehyde and ferric acid were detected on this electrode, with maximum concentrations of 43 mg L^{-1} and 18 mg L^{-1} , respectively.

The results also showed that the addition of vanadium oxide as a catalyst improved the chemical efficiency results, reaching 75% for the methane formation reaction using EDG with 20% catalyst, compared to EDG without catalyst with 5.5% efficiency. With regard to the application of the electrical charge using the same electrode, an efficiency of 38% was obtained for the methanol formation reaction, with an energy expenditure of 3.5 kWh kg^{-1} of methanol formed.

5.2 - Use of oxide EDG in ethylene oxidation

To study the oxidation of ethylene gas, EDG without a catalyst, EDG catalyzed with vanadium oxides, EDG catalyzed with palladium oxides and EDG catalyzed with silver oxides were used.

Using EDG without catalyst, ethylene glycol was quantified at 161 mg L^{-1} at 2.3 V vs. ECS and ethanol at 41 mg L^{-1} at 1.7 V vs. ECS. The variation in the concentration of the two products was different, as the increase in the applied potential led to an increase in the concentration of ethylene glycol and a decrease in the concentration of ethanol from 1.7 V vs. ECS. The maximum efficiency values for the ethylene glycol formation reaction reached 36% for chemical efficiency, 18% for electrical efficiency and the ethylene glycol formation reaction consumed a minimum of 9 kWh kg^{-1} of ethylene glycol at 2.0 V vs. ECS.

The experiments using EDG catalyzed with vanadium oxide showed promising results for the generation of ethanol, since increasing the amount of catalyst led to an increase in the formation of alcohol, reaching 98 mg L^{-1} of ethanol at 1.7 V vs. ECS with EDG catalyzed with 20% vanadium oxide, while the formation of ethylene glycol in the same EDG was 47 mg L^{-1} at 2.0 V vs. ECS. As the amount of catalyst decreased, the formation of ethanol decreased, 51 mg L^{-1} at 1.7 V vs. ECS with EDG catalyzed with 1% vanadium oxide and 104 mg L^{-1} of ethylene glycol at 2.0 V vs. ECS. These results are confirmed by the efficiency values of the ethylene glycol formation reaction, with 9% chemical efficiency in EDG with 20% catalyst and 20% efficiency in EDG with 1% catalyst. This difference was also observed in electrical efficiency, 14% for EDG with 20% catalyst and 34% for

EDG with 1% catalyst, both at 2.0 V vs. ECS. With regard to energy consumption for the ethylene glycol formation reaction, it was observed that increasing the catalyst (20%, 10% and 5%) led to an increase in energy consumption for the reaction and in the EDG with 1% catalyst, reaching approximately 37 kWh kg^{-1} of ethylene glycol.

The addition of silver oxide as a catalyst in the EDG structure showed superior results compared to vanadium catalyst and EDG without catalyst when compared to the formation of ethylene glycol. For 20% silver oxide, 195 mg L^{-1} of ethylene glycol was formed at 2.3 V vs. ECS and 16 mg L^{-1} of ethanol at 1.7 V vs. ECS. The efficiency for the ethylene glycol formation reaction showed higher values than vanadium oxide and EDG without catalyst, reaching 37% chemical efficiency, 34% electrical efficiency and 7 kWh kg^{-1} for the formation of ethylene glycol, both at 2.0 V vs. ECS.

The results with EDG catalyzed with palladium oxide showed that this catalyst was the best used for the formation of ethylene glycol, achieving the highest concentrations, efficiency values and lowest energy consumption compared to the other catalysts. In the formation of ethylene glycol, EDG catalyzed with 20% palladium oxide achieved 209 mg L^{-1} of ethylene glycol and only 12 mg L^{-1} of ethanol, showing that the addition of palladium oxide to the structure of EDG provides an increase in the formation of ethylene glycol and a decrease in the formation of ethanol. In terms of chemical efficiency, EDG catalyzed with 20% palladium showed 42% efficiency in converting ethylene into ethylene glycol. In terms of electrical efficiency, 37% efficiency was achieved, at 2.0 V vs. ECS, when applying the electrical charge to form ethylene glycol, with a minimum consumption of 6 kWh kg^{-1} of ethylene glycol formed at 2.0 V vs. ECS.

Experiments were also carried out at constant current using EDG without catalyst and catalyzed with 20% palladium. The results showed that experiments at 0.70 mA cm^{-2} showed better results, with and without catalyst, reaching 341 mg L^{-1} for EDG without catalyst and 487 mg L^{-1} for EDG with 20% vanadium oxide. In terms of chemical efficiency, the catalyzed electrodes showed better results than the electrodes without catalyst, 51% for the EDG without catalyst and 57% for the catalyzed EDG. In terms of electrical efficiency, the catalyzed EDG showed 78% efficiency and the EDG without catalyst showed 66% electrical efficiency. The energy consumption results showed that the EDG without catalyst consumed 0.9 kWh kg^{-1} and 0.4 kWh kg^{-1} for the EDG catalyzed with 20% palladium oxide.

The gas diffusion electrodes catalyzed with 20% palladium oxide were also evaluated for their stability in several consecutive experiments. In the first evaluation, a new EDG was used in three consecutive experiments lasting one hour each and the results showed that the electrode maintained the production of ethylene glycol at close values, showing that this electrode is reproducible in subsequent experiments. The amount of metals on the electrode surface was also evaluated before and after the three reproducibility experiments and the results showed small changes in the electrode surface composition, such as a decrease from 28.3% to 28.1% for ruthenium oxide and from 5.5% to 5.4% for palladium oxide. The gas diffusion electrodes showed a tendency for the concentration of ethylene glycol to stabilize as a function of time in all the experiments, which is why a four-hour experiment was carried out to assess the real stabilization profile and the results showed the profile defined for ethylene glycol from 30 minutes into the experiment.

The results of ethylene glycol production with uncatalyzed EDG and EDG catalyzed with vanadium, palladium and silver showed that palladium oxide is the best catalyst for ethylene glycol formation, with the electrode $(TiO_2)_{0.661}(RuO_2)_{0.283}(PdO_2)_{0.056}$. For the formation of ethanol, the best catalyst was vanadium oxide with the electrode $(TiO_2)_{0.66i}(RuO_2)_{0.283}(V_2O_5)_{0.056}$, but the electrodes with 20% vanadium oxide showed the worst results for the formation of ethylene glycol with 47 mg L .$^{-1}$

5.3 Continuity for future work

The results presented showed that the gaseous diffusion electrode is efficient in the selective oxidation of methane and ethylene, but it is necessary to continue the studies of this project in order to improve the results presented:

1 - Definition of the best amount of catalyst to be used in the EDG, depending on the product expected for the reaction.

2 - Study of new catalysts to be used in selective oxidation, with the aim of increasing the formation of ethylene glycol and reducing energy consumption.

3 - Study of the redox reactions of ruthenium and the catalysts used in the proposed experimental conditions

References

ANTHONY, C. R.; McELWEE-WHITE, L. Zirconia-supported phosphotungstic acid as catalyst for alkylation of phenol with benzyl alcohol. **J. Mol. Catal. A: Chem.** v.223, p.113-117, 2005.

AOKI, K.; OHMAE, M.; NANBA, T.; TAKEISHI, K.; AZUMA, N.; UENO, H.; HAYASHI, H.; UDAGAWA, N. Direct conversion of methane into methanol over MoO3/SiO2 catalyst in an excess amount of water vapor. **Catal. Today**, v.45, p.29-33, 1998.

Brazilian Chemical Industry Association. **Guide to the Brazilian Chemical Industry**. Sào Paulo: ABIQUIM, 2006, 376 p.

BEATI, A. A. G. F.; ROCHA, R. S.; OLIVEIRA, J. G.; LANZA, M. R. V. Study of ranitidine degradation via H2O2 electrogenerated/Fenton in an electrochemical reactor with gas diffusion electrodes. **Chem. Nova**, v.32, n.1, p.125-130, 2008.

BOBROVA, I. I.; BOBROV, N. N.; SIMONOVA, L. G.; PARMON, V. N. Direct catalytic of methane to formaldehyde: New investigation opportunities provided by an improved flow circulation method. **Kinet. Catal**. v.48, n.5, p.676-692, 2007.

CHEN, L. Y.; YANG, B. L.; ZHANG, X. C.; DONG, W.; CAO, K. Partial oxidation of methane to methanol in liquide phase by V2O5 catalyst. **Chin. J. Catal.** v.26, n. 11, p.1027-1030, 2005.

CHEN, L.; YANG, B.; ZHANG, X.; DONG, W.; CAO, K.; ZHANG, X. Methane Oxidation over a V2O5 catalyst in the liquid phase. **Energy Fuels**, v.20, p.915-918, 2006.

COMNINELLIS, C. Electrocatalysis in the electrochemical conversion /combustion of organic pollutants for waste water treatment. **Electrochim. Acta**, v.39, n.11/12, p.1857-1862, 1994.

COMNINELLIS, C. Electrochemical oxidation of organic pollutants for wastewater treatment. In: SEQUEIRA, C. A. C. (Ed.). **Environmental Oriented Electrochemistry**. Amsterdam: Elsevier, 1994. chap. 1, p.77-102.

COMNINELLIS, C. Theoretical model for the anodic oxidation of organic on metal oxide electrodes. **Electrochim Acta**, v.42, n.13-14, p.2009-2012, 1997.

FAJARDO, C. A. G.; NIZNANSKY, D.; N'GUYEN, Y.; COURSON, C.; ROGER, A. C. Methane selective oxidation to formaldehyde with Fe-catalysts supported in silica or incorporated into the support. **Catal. Commun.** v.9, n.864-869, 2008.

FORTI, J. C.; OLIVI, P.; ANDRADE, A. R. Characterization of DSA®-type coating with nominal composition Ti/Ru0,3Ti$_{(0,7-x)}$SnxO2 prepared via a polymeric precursor. **Electrochim. Acta**, v.47, p.913-920, 2001.

FORTI, J. C.; ROCHA, R. S.; LANZA, M. R. V.; BERTAZZOLI, R. Electrochemical synthesis of hydrogen peroxide on oxygen-fed graphite/PTFE electrodes modified by 2-ethylanthraquinone. **J. Electroanal. Chem,** v.601, p.63-67, 2007.

FOTI, G. GANDINI, D. COMNINELLIS, C. Oxidation of organics by intermediates of water discharge on IrO2 and synthetic diamond anodes. **Electrochem Solid-State Lett**. v.2, n.5, p.228230, 1999.

FOULDS, G. A.; GRAY, B. F. Homogeneous gas-phase partial oxidation of methane to methanol and formaldehyde. **Fuel Process. Technol.** v.42, p.129-150, 1995.

GANG, X.; BIRCH, H.; ZHU, Y.; HJULER, H. A.; BJERRUM, N. J. Direct oxidation of methane to methanol by mercuric sulfate catalyst. **J. Catal.** v.196, n.287-292, 2000.

GHARIBI, H.; ZHIANI, M.; ENTEZAMI, A. A.; MIRZAINE, R. A.; KHEIRMAND, M.; KAKAEI, K. Study of polyaniline doped with trifluoromethane sulfonic acid in gas-diffusion electrodes for proton-exchange membrane fuel cells. **J. Power Sources**. v.155, p.138-144, 2006.

INDARTO, A. A review of Direct Methane Conversion to Methanol by Dielectric Barrier Discharge. **Transactions on Dielectrics and Electrical Insulation**, v.15, n.4, p.1038-1043, 2008.

KHOKHAR, M. D.; SHUKLA, R. S.; JASRA, R. V. Selective oxidation of methane by molecular oxygen catalyzed by a bridged binuclear ruthenium complex at pressures and ambient temperature. **J. Mol. Catal. A: Chem**, v.299, p.108-116, 2009.

L. A. de Faria, J. F. C. Boodts, S. Trassati, Physico-chemical and electrochemical characterization of Ru-based ternary oxides containing Ti and Ce. **Electrochim. Acta**, v.37, p.2511-2518, 1992.

LU, G.; SHEN, S.; WANG, R. Direct oxidation of methane to methanol at atmospheric pressure in CMR and RSCMR. **Catal. Today**, v.30, p.41-48, 1996.

MICHALKIEWICZ, B. Methane Conversion to methanol in condensed phase. **Kinet. Catal.** v.44, n.6, p.874-878, 2003.

MIWA, D. W.; MALPASS, G. R. P.; MACHADO, S. A. S.; MOTHEO, A. J. Electrochemical degradation of carbaryl on oxide electrodes. **Water Resour.** v. 40, p.3281-3289, 2006.

O. Simond, V. Schaller, Ch. Comninellis, Theoretical model for the anodic oxidation of organics on metal oxide electrodes. **Electrochim. Acta**, v.42, p.2009, 1997.

PANTU, P.; GAVALAS, G.; R. Methane partial oxidation on Pt/CeO2 and Pt/Al2O3 catalysts. **Appl. Catal. A**, v.223, p.253-260, 2002.

PARK, E.; CHOI, S. H.; LEE, J. S. Characterization of Pd/C and Cu catalysts for the oxidation of methane to a methanol derivative. **J. Catal.** v.194, p.33-44, 2000.

PELEGRINO, R. R. L.; VICENTIN, L. C.; DE ANDRADE, A. R.; BERTAZZOLI, R. Thirty minutes laser calcination method for the preparation of DAS® type oxide electrodes. **Electrochem. Commun.** v.4, p.139-142, 2002.

RAJA, R.; RATNASAMY, P. Direct conversion of methane to methanol. **Appl. Catal. A**, v.158, p.L7-L15, 1997.

RAZUMOVKY, S. D.; EFREMENKO, E. N.; MAKHLIS, T. A.; SENKO, O. V.; BIKHOVSKY, M. Y.; PODMASTER, V. V.; VARFOLOMEEV, S. D. effect of immobilization on the main dynamic characteristics of the enzymatic oxidation of the methane to methanol by bacteria *Methylosinus sporium* B-2121. **Russ. Chem. Bull.** v.57, n.8, p.1633-1636, 2008.

ROCHA, R. S.; BEATI, A. A. G. F.; OLIVEIRA, J. G.; LANZA, M. R. V. Evaluation of the degradation of diclofenac sodium using 1 H'l- fenton in an electrochemical reactor. **Quim Nova**, v.32, n.2, p.254-258, 2008.

SHEVERDENKIN, E. V.; ARUTYUNOV, V. S.; RUDAKOV, V. M.; SAVCHENKO, V. I.; SOKOLOV, O. V. Kinetics of partial oxidation of alkanes at high pressures: oxidation of ethane and methane-ethane mixtures. **Theor. Found. Chem. Eng.** v.38, n.3, p.311-315, 2004.

SOEHN, M.; LEBERT, M.; WIRTH, T.; HOFMANN, S.; NICOLOSO, N. Design of gas diffusion electrodes using nanocarbon. **J. Power Sources**. v.176, p.494-498, 2008.

TABATA, K.; OKURA, I. Hydrogen and Methanol formation Utilizing Bioprocesses. **Journal of the Japan Petroleum Institute**, v.51, n.5, p.255-263, 2008.

TRASSATI, S. Electrocatalysis: understanding the success of DSA® . **Electrochim. Acta**, v.45, p.2377, 2000.

WANG, C. B.; HERMAN, R. G.; SHI, C.; SUN, Q.; ROBERTS, J. E. V2O5-SiO2 xerogels for methane oxidation to oxygenates: preparation, characterization, and catalytic properties. **Appl. Catal., A**, v.247, p. 321-333, 2003.

WANG, Y.; OTSUKA, K. Catalytic oxidation of methane to methanol with H2-O2 gas mixture at atmospheric pressure. **J. Catal.** v.155, p.256-267, 1995.

www.abiquim.org.br, accessed on 07/08/2008 at 14:30 h.

www.gamagases.com.br, accessed on 10/03/2010 at 19:55 h.

www.metanor.org.br, accessed on 07/08/2008 at 17:00.

www.sigmaaldrich.com, accessed on 07/08/2008 at 14:00.

ZHANG, Q.; HE, D.; HAN, Z.; ZHANG, X.; ZHU, Q. Controlled partial oxidation of methane to methanol/formaldehyde over Mo-V-Cr-Bi-Si oxide catalysts. **Fuel**, v.81, p.1599-1603, 2002.

Printed by Books on Demand GmbH, Norderstedt / Germany